THERMAL EXPANSION 7

THERMAL EXPANSION 7

Edited by

David C. Larsen

IIT Research Institute
Chicago, Illinois

PLENUM PRESS · NEW YORK AND LONDON

Library of Congress Cataloging in Publication Data

International Thermal Expansion Symposium
  (7th: 1979: Chicago, Ill.)
  Thermal expansion 7.

  ''Proceedings of the Seventh International Thermal Expansion Symposium, hosted
and sponsored by the IIT Research Institute, held concurrently with the Sixteenth International Thermal Conductivity Conference, November 7–9, 1979, in Chicago, Illinois'' — Verso t.p.
  Includes bibliographical references and index.
  1. Expansion (Heat) — Congresses. I. Larsen, David C. II. IIT Research Institute. III.
Title. IV. Title: Thermal expansion seven.

QC281.5.E9I58 1979   536'.41   82-9083
ISBN 978-1-4684-8269-0   ISBN 978-1-4684-8267-6 (eBook)   AACR2
DOI 10.1007/978-1-4684-8267-6

Proceedings of the Seventh International Thermal Expansion
Symposium, hosted and sponsored by the IIT Research Institute,
held concurrently with the Sixteenth International Thermal
Conductivity Conference, November 7–9, 1979, in Chicago, Illinois

Plenum Press, New York
a Division of Plenum Publishing Corporation
233 Spring Street, New York, N.Y. 10013

FOREWORD

The International Thermal Expansion Symposium was started in
1968 with the initiative of Messrs. R. K. Kirby and P. S. Gaal.
These Symposia cover the developments and advances in theoretical
and experimental studies of the thermal expansion of solids and its
relation to other related properties, and provide a broadly based
forum for researchers actively working in this field to convene on
a regular basis to exchange their ideas and experiences and report
their findings and results.

The Symposia have been self-perpetuating and are an example of
how a technical community with a common purpose can transcend the
invisible artificial barriers between disciplines and gather together in increasing numbers without the need of national publicity and
continuing funding support, when they see something worthwhile going
on.

Of the first five Symposia only three published formal Proceedings. However, effective with the Sixth Symposium in 1977
when our Center for Information and Numerical Data Analysis and
Synthesis (CINDAS) of Purdue University became the permanent Spon-
sor of the Symposia, a policy of publishing formal Proceedings on
a continuing and uniform basis has been established. Thus, including the present volume, the following formal Proceedings have been
published:

Symposium and Year	Title of Volume	Publisher and Year
2nd (1970)	SYMPOSIUM ON THERMAL EXPANSION OF SOLIDS, Journal of Applied Physics, 41 (13), pp. 5043-5154	American Institute of Physics (1970)
3rd (1971)	THERMAL EXPANSION - 1971 AIP Conference Proceedings No. 3	American Institute of Physics (1972)
4th (1973)	THERMAL EXPANSION - 1973 AIP Conference Proceedings No. 17	American Institute of Physics (1974)
6th (1977)	THERMAL EXPANSION 6	Plenum Press (1978)
7th (1979)	THERMAL EXPANSION 7	Plenum Press (1982)

Mr. David C. Larsen, General Chairman of the Seventh Symposium is to be congratulated for his painstaking efforts which made the present volume possible. CINDAS looks forward to working with future host institutions to ensure that future Symposia continue to produce high-quality volumes of Proceedings in this important, specialized field.

This Foreword should have been written by Dr. Y. S. Touloukian, the founder and founding Director of CINDAS for the past 25 years. It was all owing to Dr. Touloukian's great efforts that the Proceedings of recent and future Symposia have been and will be published formally by Plenum Press on a continuing and uniform basis, thus making them a continuing series of uniform volumes serving as the permanent major vehicle for the reporting of research results on thermal expansion. I regret most deeply to report that Dr. Touloukian died suddenly on 12 June 1981. His passing away is a great loss to the field of thermophysical properties, to which he had devoted his entire life.

C. Y. Ho
Interim Director
Center for Information and
   Numerical Data Analysis
   and Synthesis
Purdue University

November, 1981
West Lafayette, Indiana

PREFACE

The 7th International Thermal Expansion Symposium (ITES) was
hosted and sponsored by IIT Research Institute, Chicago, Illinois
on November 7-9, 1979, and was held concurrently with the 16th
International Thermal Conductivity Conference (ITCC). The General
Chairman was David C. Larsen. A list of previous ITES meetings in
the series is shown in the accompanying table.

The 7th ITES and 16th ITCC were attended by 113 people, re-
presenting 11 countries. When co-authors are included in the tabu-
lation, 16 countries are represented, reflecting the truely international community of thermophysicists contributing to the ITCC/ITES.
Over 90 papers were presented at the joint meetings, arranged in
thirteen ITCC and five ITES technical sessions, and two plenary
sessions. Eighteen ITES manuscripts are compiled herewith in the
ITES conference proceedings (the ITCC proceedings are also available
from Plenum). Predominant subjects of the ITES include 1) high and
low temperature measurements, 2) measurement techniques, 3) results
on composite materials, 4) theory and correlations, and 5) applications.

A highlight of the 7th ITES/16th ITCC meetings in Chicago was
the dinner banquet. K. D. Maglic outlined the plans for the 7th
European Thermophysical Properties Conference, held in Antwerp,
Belgium in June 1980. A. Cezairliyan overviewed the purpose and
scope of the International Thermophysics Congress. Additionally,
T. A. Hahn told of plans for the Eighth ITES.

The 7th ITES is indebted to the Session Chairmen for conducting
the technical sessions and reviewing manuscripts for inclusion in
the conference proceedings. The support, direction, and advice
given by the ITES Governing Board throughout this project is greatly
appreciated. Thanks are also due CINDAS/Purdue University for
co-sponsorship of the ITES, in the form of editorial guidance, and
interaction with the publisher of this volume. Locally, the efforts
of IITRI's Special Events Coordinators, C. J. Galassi and D. Lancaster,
in organizing and running the 3-day meeting are considered invaluable.
The work of secretaries D. J. Dickson and Y. Bradley in typing
announcements, programs, and various manuscripts is hereby acknow-
ledged. Thanks are also due Y. Harada for help in conducting
the meeting and preparing the proceedings volume. And finally, a
very special thank you is extended to J. W. Adams for continual
assistance, support, and encouragement throughout this project.

<u>PREVIOUS THERMAL EXPANSION SYMPOSIA</u>

Conf.	Year	Host Organization and Site	Chairman
1	1968	U.S. National Bureau of Standards and Westinghouse Astronuclear Lab (Gaithersburg, MD)	R.K. Kirby P.S. Gaal
2	1970	University of Illinois (MRL) and Sandia Laboratories (Santa Fe, NM)	R.O. Simmons D.C. Wallace
3	1971	Corning Glass Works and University of Toronto (Corning, NY)	H. Hagy G.M. Graham
4	1973	Purdue University and U.S. Air Force Materials Laboratory (Lake of the Ozarks, MO)	R.E. Taylor G.L. Denman
5	1975	Oak Ridge National Laboratory University of Connecticut (Storrs, CT)	T.G. Godfrey P.G. Klemens
6	1977	Atomic Energy of Canada Limited (Hecla Island, Manitoba, Canada)	I.D. Peggs
7	1979	IIT Research Institute (Chicago, IL)	D.C. Larsen

The next meeting in this series was held in June 1981 at the National Bureau of Standards in Gaithersburg, Maryland. The International Joint Conferences on Thermophysical Properties featured the Eighth ASME Symposium on Thermophysical Properties (J. V. Sengers, Chairman), the Seventeenth International Thermal Conductivity Conference (J. G. Hust, Chairman), and the Eighth International Thermal Expansion Symposium (T. A. Hahn, Chairman). These three concurrent conferences were coordinated by the International Thermophysics Congress (A. Cezairliyan, Chairman).

Chicago, Illinois   David C. Larsen
October, 1981   General Chairman, 7th ITES

CONTENTS

SESSION 1

HIGH AND LOW TEMPERATURE MEASUREMENTS

Chairman: P. Wagner
Los Alamos Scientific Laboratory

SESSION 2

MEASUREMENT TECHNIQUES

Chairman: T.A. Hahn
U.S. National Bureau of Standards

SESSION 6

MISCELLANEOUS MATERIALS/APPLICATIONS

Chairman: W.A. Plummer
Corning Glass Works

ITES SESSION 1: HIGH AND LOW TEMPERATURE MEASUREMENTS

Session Chairman: P. Wagner
                      Los Alamos Scientific Laboratory
                      Los Alamos, NM

MEASUREMENT OF THERMAL EXPANSION AND VOLUME CHANGES IN

PARTIALLY STABILIZED ZIRCONIA

W. A. Plummer and S. T. Gulati

Research and Development Laboratories
Corning Glass Works
Corning, N. Y. 14830

ABSTRACT

One of the most promising applications of partially stabilized zirconia is in the metal extrusion industry in which CaO or MgO stabilized ZrO_2 dies have outperform- ed the conventional metal dies. The secret of this success lies in the polyphase nature of PSZ and its high toughness, low expansion and excellent thermal shock re- sistance which are linked to volume changes associated with phase transformation during heating and cooling.

In this paper we report the thermal expansion data for three different versions of MgO stabilized zirconia in the temperature range 25-1300°C. The densification associated with monoclinic to tetragonal transformation during heating and expansion due to reversion of the phases during cooling can be estimated from expansion curves as can the transformation temperature. The vol- ume change data thus obtained are a function of the monoclinic content and firing conditions and are in good agreement with literature values. The expansion coef- ficient and volume changes show appreciable difference during remeasurement due to microcracking associated with phase changes. The changes in Young's Modulus with temperature are also found to be consistent with density changes during heating.

INTRODUCTION

Partially stabilized zirconia (PSZ) has found successful application as a die material for extruding

3

ferrous and nonferrous products, notably rods and
tubing.[1,2]   Extrusions are made at pressures approaching
1 GPa, temperatures as high as 1250°C and speeds of 20
meters per second. Conventional die materials such as
tool steels, tungsten carbide and super alloys such as
stellite undergo severe deformation under these conditions and result in poor dimensional control as well as
poor surface quality. This necessitates frequent remachining which, in turn, requires large die inventories
and excessive down time for die changes. A further dif-
ficulty arising from conventional dies is that the life
of the die cases and holders is shortened by the high
temperatures they reach due to the high conductivity of
metal dies.

A potential replacement is a disposable low-cost
zirconia die. Partially stabilized zirconia exhibits
high strength, high fracture toughness, excellent thermal stability and low creep; it has low conductivity,
non-wetting property and takes a high polish.[3,5]   These
properties make it a good candidate for the extrusion
die that could out-perform conventional dies by giving
more extrusions per die without requiring remachining.
Hence production losses due to downtime are minimized,
if not eliminated altogether.

In this paper we report the thermal expansion data
and the volume changes associated with phase transformation for three partially stabilized (MgO stabilizer)
zirconia compositions. These properties are needed i)
to compute thermal stresses during extrusion, ii) to
assess the useful temperature range of each material and
iii) to arrive at the optimum die geometry for a given
application.

In a recent paper Gulati et al.[1] used the thermal
and mechanical properties for determining stress levels
at extrusion pressure and temperature. A preliminary
analysis showed that the strength of PSZ is inadequate
to combat these stresses. However, by shrink fitting
the dies into steel die cases sufficient compressive
stresses can be attained to withstand extrusion stresses.
A 0.25 mm diametral interference in a 5 cm die, for example, gives strengths in excess of 1 GPa. One limita-
tion of the shrink-fit technique is that the case-hard-
ened die holder cannot be heated to above 650°C. Secondly, the out-of-round tolerance in PSZ die can cause
severe tensile stresses during die insertion and lead
to premature failure. Using the finite element method,
stresses due to extrusion pressure, temperature and

volume change were shown to approach 1 GPa at the extrusion surface. With a properly designed and encased PSZ die, field trials gave up to 20 extrusions per die of ferrous rod with excellent dimensional control and surface finish.

PARTIALLY STABILIZED ZIRCONIA

Zirconia exists in a monoclinic structure at ambient temperature and pressure. At 1200°C and atmospheric pressure it undergoes transformation to a tetragonal structure. This transformation is accomplished by a large volume change of ~3.25%. The addition of certain oxides, MgO, CaO, Y_2O_3, for example, results in the formation of a stable cubic phase. The thermal expansion of a fully stabilized body is 12×10^{-6} / C which together with the low thermal conductivity results in poor thermal shock resistance. By controlling the stabilizer content, a partially stabilized material can be formed which consists of cubic and monoclinic phases. The thermal expansion of such a partially stabilized body is only slightly higher than that of monoclinic material and the volume change is significantly reduced, thus enhancing the thermal shock resistance. Other factors contributing to the thermal shock resistance are (1) microcracking caused by stresses due to transformation from tetragonal to monoclinic phase, (2) dissipation of strain energy by intragranular precipitates at unstabilized ZrO_2, and (3) existence of metastable tetragonal phase with a grain size of ≃100 nm.

The three materials discussed in this paper, ZIRCOA®* compositions 2032, 1027 and 2016, are MgO stabilized ZrO_2 and contain 30%, 40% and 50-60% monoclinic phase respectively as determined by X-ray diffraction analysis. In the first two materials nearly all of the monoclinic phase is in intragranular precipitates - while in composition 2016 the monoclinic phase is approximately evenly distributed between inter and intragranular phases. This exhibits the best thermal shock resistance.

PHYSICAL PROPERTIES

Young's Modulus

We include this property since its change with temperature is consistent with density changes associated

*ZIRCOA® Products, Ceramic Products Division, Corning Glass Works.

with phase transformation. Young's modulus was measured by the sonic resonance technique. Rectangular bars ground to precise dimensions were suspended in a furnace using asbestos fiber. The fundamental resonant frequency in the flexural mode was measured as function of temperature during heating at a rate of 160°C/hour. The modulus was calculated from the expression[6]

$$E = 9.45 \times 10^{-13} \left(\frac{L^3}{bt^3}\right) T_1 Mf^2 \qquad (1)$$

where

    L = length of bar, cm
    b = width of bar, cm
    t = thickness of bar, cm
    M = mass of bar, gm
    f = resonance frequency, Hz
    T_1= correction factor for fundamental flexural mode
       $\approx$ 1.1 for specimens used

The results for all three compositions are shown in Fig. 1. There is an initial decrease in modulus to about

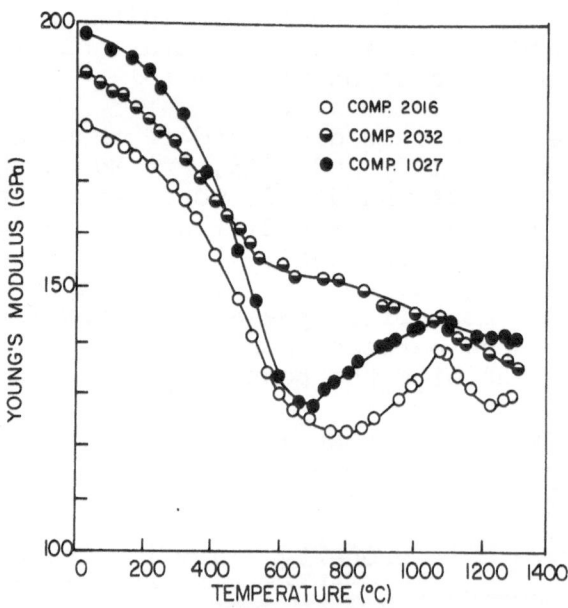

Fig. 1 Young's modulus of compositions 1027, 2016 and 2032 as function of temperature (heating only).

600°C. Between 600°C and 800°C this decrease is either
retarded or reversed. Composition 2016 shows a sharp
maximum at about 1100°C whereas comp. 1027 shows only a
slight maximum. Composition 2032 exhibits rapid decrease
in modulus to 500°C followed by a gradual decrease to
1300°C. The initial decrease in modulus is the normal
temperature coefficient. The changes at 800°C and
higher are associated with the phase transition from
monoclinic to tetragonal structure.

Thermal Expansion

     Thermal expansion was measured from 25°C to 1300°C
using a Theta alumina dilatometer at a heating rate of
260°/hour. Measurements were made during heating and
cooling. In addition the thermal expansion of composition 1027 was determined from -196°C to 25°C using a
silica rod-type dilatometer. A fully stabilized zirconia
has also been measured.[7] A composite curve for pure
ZrO_2 , and partially and fully stabilized zirconias is
shown in Fig. 2. These data are under heating conditions

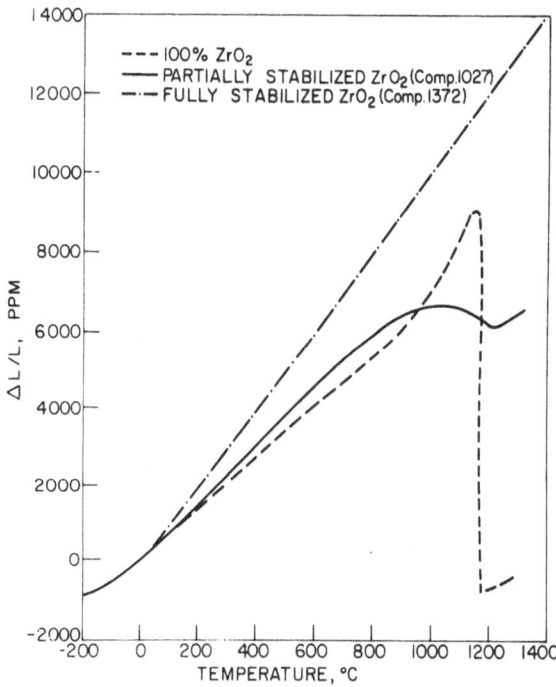

Fig. 2   Thermal expansion curves for fully stabilized,
         partially stabilized and pure zirconia (heating
         only).

and the expansion of the "pure" material is taken from
Fehrenbacher and Jacobson.[7] The initial expansion of
ZrO_2 is lower than that of either the partially or fully
stabilized zirconia. At about 800°C the expansivity
increases. Then at approximately 1200°C the material
undergoes conversion from the monoclinic to tetragonal
phase. This phase change is accompanied by a decrease
in length of 10,000 ppm which corresponds to a 3% volume
change. The highest expansion is exhibited by the fully
stabilized material with accompanying poor thermal shock
resistance. The expansion of the partially stabilized
zirconia is intermediate and its volume change during
transformation is markedly lower. The magnitude of this
volume change is dependent upon the proportion of monoclinic and cubic phases and on the size and distribution
of the monoclinic phase.

The three compositions studied illustrate differences due to composition and microstructure. The latter
is affected by thermal history. The composition 2032
contains 30% monoclinic phase nearly all of which is in
intragranular precipitates. This exhibits little apparent volume change on heating as shown in Fig. 3. There
is a slight expansion, however, at 600°C on cooling.

Composition 1027 has about 40% monoclinic phase
which is present as intragranular precipitates. The
expansion curve during first cycle, Fig. 4, shows a
change in expansion at 600°C and again at 1100°C during
heating. However, only one expansion change is exhibited during cooling. In the second cycle, Fig. 5, only
the 1100°C volume change is observed during heating.
The increase in volume on cooling occurs at the same
temperature as for the first cycle but the magnitude is
somewhat larger. This could be due to an incomplete
thermal history of the initial sample.

The third material, composition 2016, exhibits significant differences between the first cycle, Fig. 6,
and second cycle, Fig. 7. On second heating the volume
change is smaller. However, the volume changes on cooling are similar resulting in substantial hysteresis.
This is partly attributed to the significant microcracking which composition 2016 is known to exhibit due to
the addition of small amounts of SiO_2.

In reporting the average expansion for such a material, the normal practice of expressing the average coefficients relative to 25°C is not helpful. The values
so measured would reflect the volume changes associated

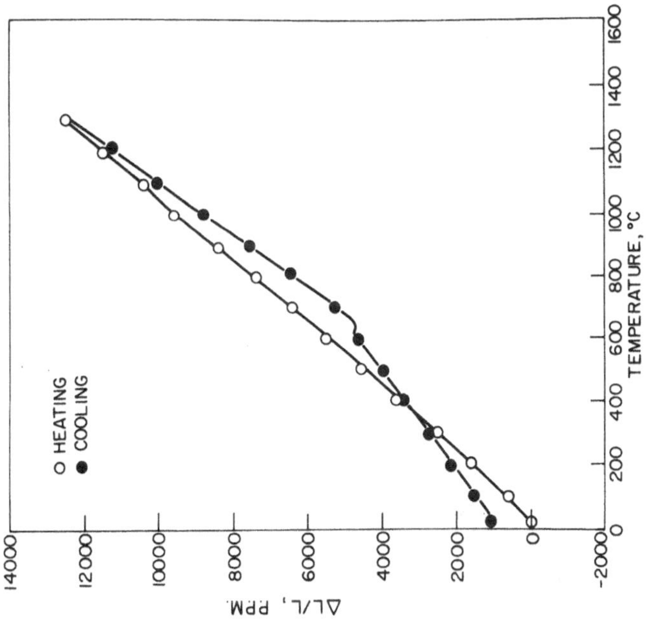

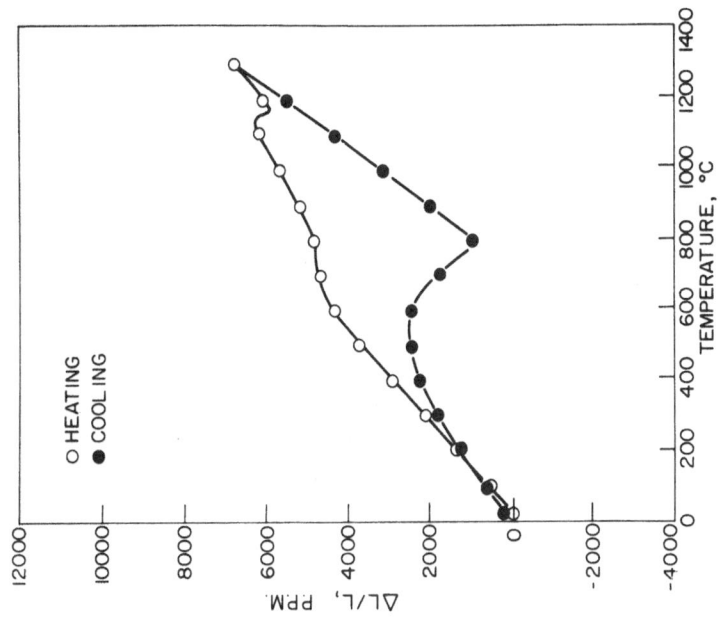

Fig. 4   Thermal expansion curve for
composition 1027 (first cycle).

Fig. 3   Thermal expansion curve for
composition 2032.

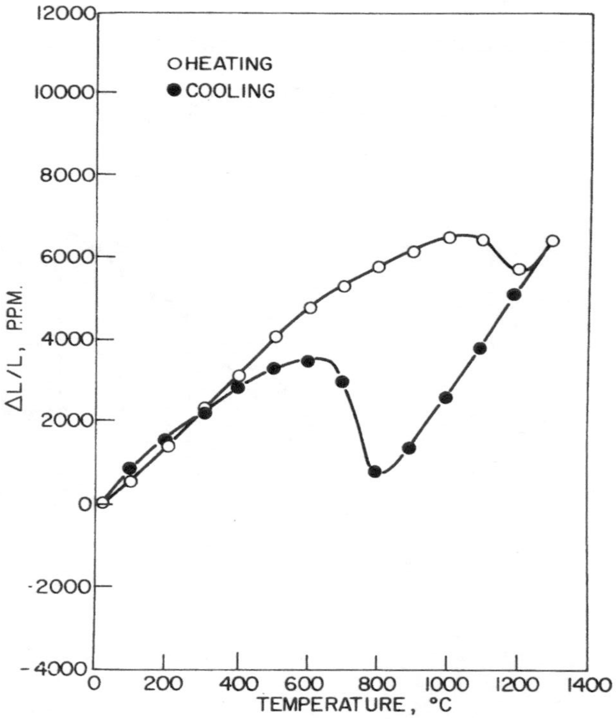

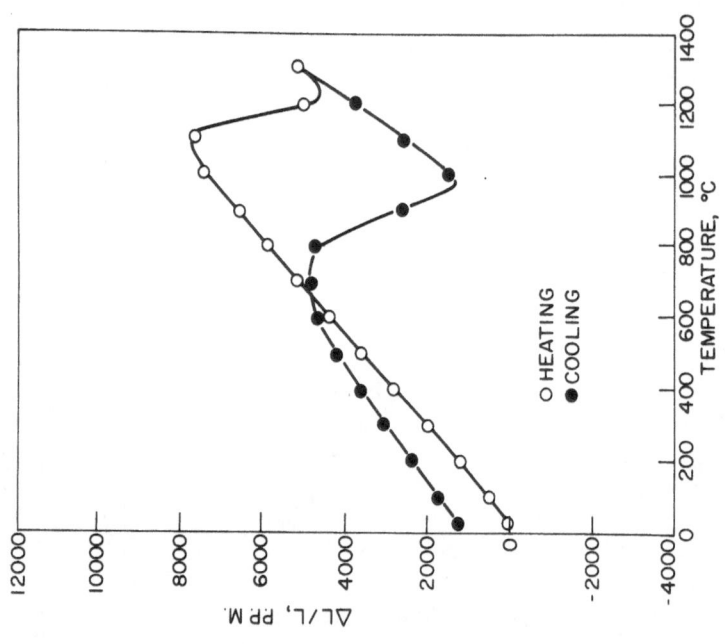

Fig. 6   Thermal expansion curve for
composition 2016 (first cycle).

Fig. 5   Thermal expansion curve for
composition 1027 (second cycle).

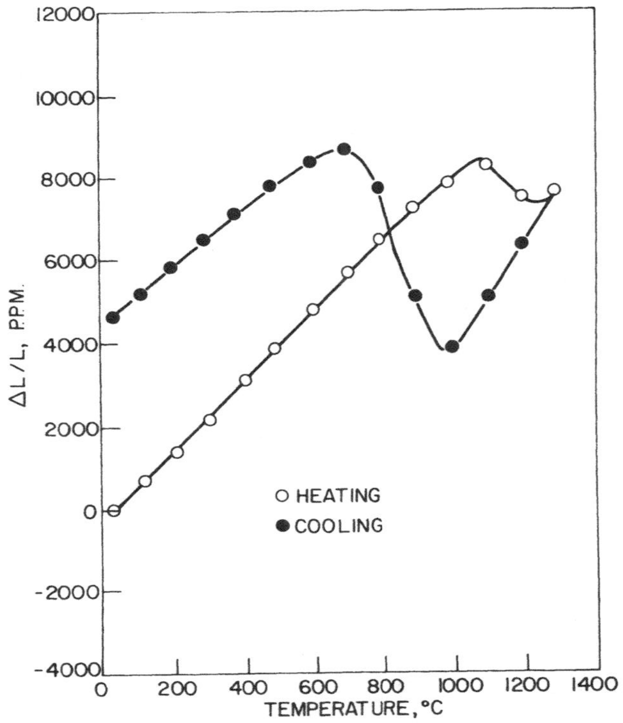

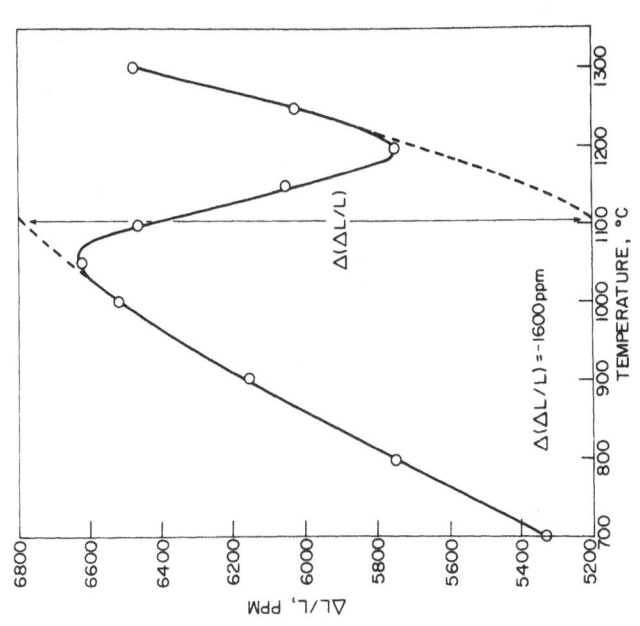

Fig. 8   Estimating volume change from length change; comp. 1027, second cycle.

Fig. 7   Thermal expansion curve for composition 2016 (second cycle).

with transformation rather than the thermal expansion
per se. A better method is to report the average coefficient above and below the transition range. The temperature intervals chosen were 25°C to 500°C and 1200°C
to 1300°C. The values calculated in this manner are
shown in Table I.

TABLE I.   THERMAL EXPANSION OF
PARTIALLY STABILIZED ZIRCONIA

MATERIAL	AVERAGE EXPANSION COEFF; $10^{-6}/°C$			
	25 - 500°C		1200 - 1300°C	
	HEATING	COOLING	HEATING	COOLING
COMP. 1027	7.7	4.7	7.0	13.0
COMP. 1027(RERUN)	8.5	6.0	7.3	13.7
COMP. 2016	8.0	6.0	5.0	13.3
COMP. 2016(RERUN)	8.4	6.5	5.5	13.1
COMP. 2032	9.5	6.0	10.8	12.7

Volume Change

The volume changes accompanying the monoclinic-
cubic transition are estimates based on the changes in
ΔL/L values in the transformation region. These estimates can be obtained in a number of ways. The most
obvious and direct method is to extrapolate the expansion curves below and above the transformation range
and calculate the difference in length at the mid-point
of the transformation region. Figure 8 illustrates this
method for the second run on composition 1027. A value
of -1600 ppm is found for ΔL/L at 1100°C.

A second method is to calculate and plot the average coefficient vs. temperature. In the example chosen
(Figure 9) we have selected 600°C as the reference temperature for calculating the coefficient. At 1100 C
the change in $\bar{\alpha}$ is -3.35 x $10^{-6}/°C$. Multiplying this
value by the temperature difference gives -1760 ppm. A
third method is to plot the instantaneous coefficient
as function of temperature. The change in length will
be the area under the curve between 900° and 1200°C,
Fig. 10. This method gave a value of -1585 ppm. The
mean value of ΔL/L is -1620 ± 50 ppm. The values calculated for ΔV/V are therefore accurate to ± 3%.

The volume changes for all three compositions were
calculated by one or more of the above methods and are
summarized in Table II. Their average values, notably

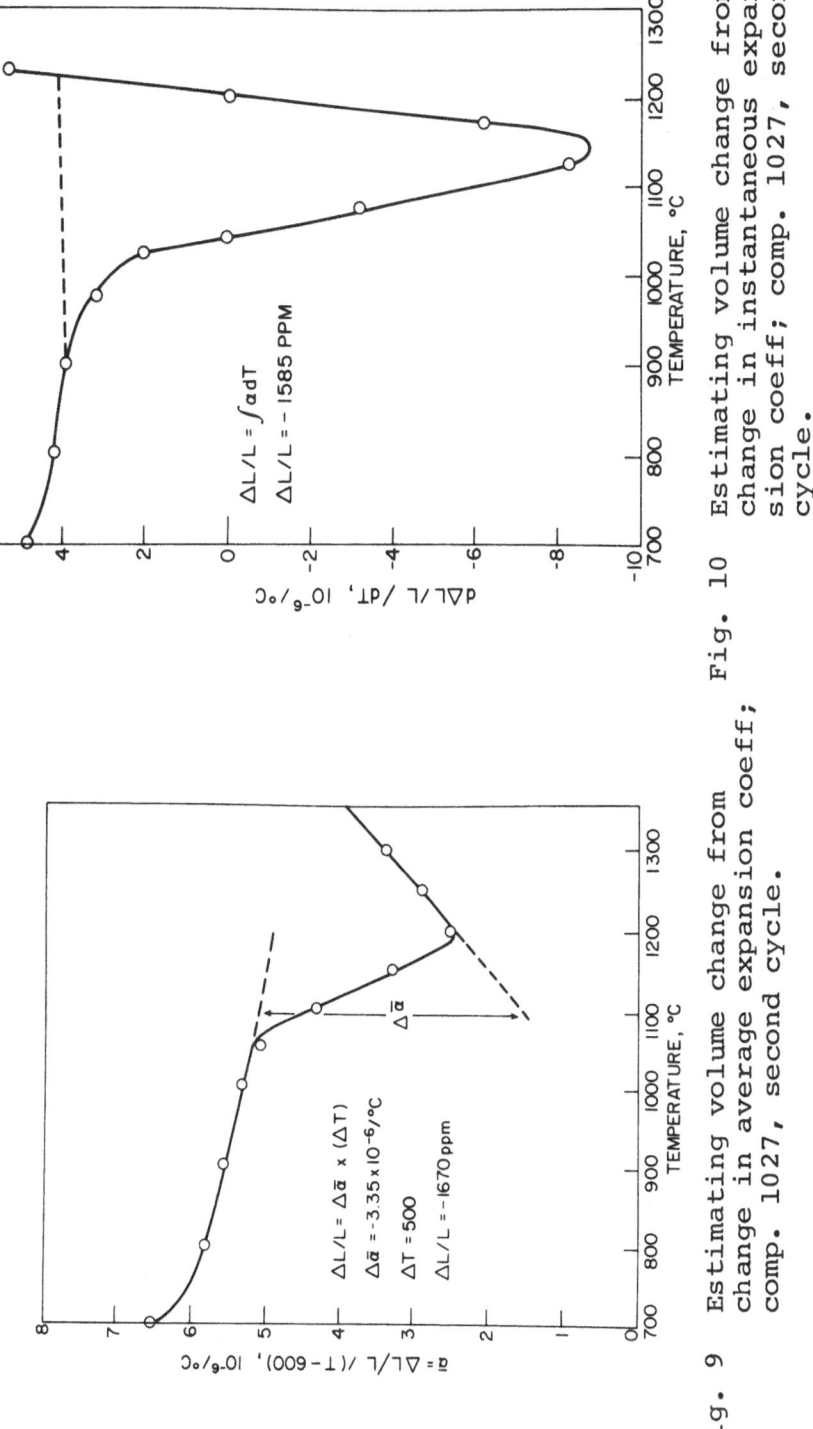

Fig. 10 Estimating volume change from change in instantaneous expansion coeff; comp. 1027, second cycle.

Fig. 9 Estimating volume change from change in average expansion coeff; comp. 1027, second cycle.

TABLE II.   VOLUME CHANGES IN
PARTIALLY STABILIZED ZIRCONIA

MATERIAL	$\Delta V/V$	
	HEATING	COOLING
COMP. 1027	0.0027	0.0087
COMP. 1027 (RERUN)	0.0048	0.0129
COMP. 2016	0.0100	0.0145
COMP. 2016 (RERUN)	0.0100	0.0170
COMP. 2032	0.0006	0.0010

during cooling, are consistent with % monoclinic content
and agree fairly well with literature data.[8-10]

The transformation temperature discussed above
refers to ambient pressure. Kulcinski[3] has reported
the change in transformation temperature with pressure.
This is important for two reasons. First, as noted in
the introduction, during extrusion the die is subjected
to pressures approaching 1 GPa. This can alter the
temperature at which the transition occurs during extrusion. It is also the reason why microstructure can be
so important in determining the magnitude of volume
changes which occur. Garvie[2] postulates that for gran-
ular sizes 100 nm or less local stresses are high enough
to inhibit the transformation to the monoclinic phase.
Hence a partially stabilized material can consist of a
dispersion of metastable tetragonal phase in the stabilized cubic phase. Such a material would not exhibit
volume changes and hence would be significantly stronger
(due to absence of microcracking). On the other hand,
its fracture toughness would be reduced significantly
making it more prone to catastrophic failure.

CONCLUSIONS

Thermal expansion and volume changes have been
measured for three compositions of partially-stabilized
zirconia. These measurements were part of a program to
characterize the material as a candidate for extrusion
dies. The volume changes which occur are not only a
function of the amount of stabilized cubic phase, but
also of the microstructure and thermal history of the
specimen. The size and distribution of the monoclinic
phase have much influence on the degree of transformation between monoclinic and tetragonal structures, and
on the amount of microcracking during cycling. These
effects can be used to explain differences in physical

properties during repeat measurements. The dependence of Young's modulus on temperature has also been reported and found to be consistent with structural changes during phase transformation.

REFERENCES

1. S. T. Gulati, J. D. Helfinstine and A. D. Davis; J. Am. Ceram. Soc. (to appear).
2. R. C. Garvie, R. H. Hannink and R. T. Pascoe; Nature (London), 258 (553) pp. 703-704 (1975).
3. G. L. Kulcinski; J. Am. Ceram. Soc. 51 (10), pp. 582-584, (1968).
4. D. L. Porter and A. H. Heuer; J. Am. Ceram. Soc. 60 (3-4), pp. 183-184 (1977).
5. T. K. Gupta; Fracture Mechanics of Ceramics, ed. R. C. Bradt, D. P. H. Hasselman and F. F. Lange; Plenum Press (1978).
6. S. Spinner and W. E. Tefft; ASTM Proc, 61, pp. 1221-38 (1961).
7. L. L. Fehrenbacher and L. A. Jacobson; J. Am. Ceram. Soc. 48 (3), pp. 157-161 (1965).
8. S. M. Lang; J. Am. Ceram. Soc. 47 (12), pp. 641-644 (1964).
9. T. H. Nielsen and M. H. Leipold; J. Am. Ceram. Soc. 47 (3), p. 155 (1964).
10. D. M. Shakhtin, E. V. Levintovich, T. L. Pivovar and G. G. Eleseeva; Akademii Nauk SSSR, Neorganicheskie Materialy, 4 (9), pp. 1603-1604 (1967).

THERMAL EXPANSION OF NICKEL TO 2300 K*

W. D. Drotning

Sandia National Laboratories**

Albuquerque, New Mexico 87185

ABSTRACT

The thermal expansion of nickel was measured from ambient temperature to 2300 K using the gamma attenuation technique. The high temperature measurement system uses a high-vacuum tungsten mesh furnace designed for passage of a collimated gamma beam through the sample region. The gamma counting system incorporates a thermostatically-controlled detector and automatic gain control to achieve count rate stability. A precision analysis is given for the use of the technique to measure changes in molten density as a function of temperature, and a discussion of possible systematic errors in the use of the measurement technique is presented. The linear thermal expansion of solid nickel from ambient to the melting point (1728 K) was observed to be slightly larger than provisional values suggested in the literature. The thermal expansion of molten nickel was measured from melt to 2300 K. The volumetric expansion coefficient was found to be $-(7.35 \pm 0.09) \times 10^{-4}$ g/cm^3 K over this temperature range, which is compared with measurements by previous investigations which range from -6.4 to -12.1×10^{-4} g/cm^3 K.

INTRODUCTION

The use of the gamma attenuation technique for thermal expansion measurements of molten materials has been discussed previously.[1,2] However, very little data has been obtained using the

_____

*This work was supported by the U.S. Department of Energy (DOE), under Contract DE-AC04-76-DP00789.
**A U.S. DOE facility.

technique in the very high temperature regime, where the method becomes particularly advantageous over other techniques. In this paper, we present the first data obtained using the gamma attenuation method on molten nickel to 2300 K. Data from previous investigators on the expansion of molten nickel differ by amounts much greater than the experimental precision would allow. Our data differ significantly from the expansion recommended in the most recent review articles on the subject.

Recent work has shown that the technique can also be applied to the measurement of the thermal expansion of isotropic solids.[3] Here we also report data on the thermal expansion of polycrystalline nickel from ambient to the melting point.

The following section of this paper describes the gamma attenuation technique data analysis methods for the thermal expansion of isotropic solids as well as the density of molten materials. A precision analysis is included for the measurement of changes in density of molten materials. The next section describes the experimental apparatus used to extend the temperature measurement range of the gamma attenuation technique to 3000 K. Finally, the results of measurements on solid and molten nickel are presented and discussed.

ANALYTICAL METHODS

The density of a material may be determined from the measured attenuation of a collimated gamma ray beam, as described by

$$I(T) = I_o(T) \exp\left[-u\rho(T)\ell(T)\right] \quad . \tag{1}$$

The gamma intensities I and I_o refer to the beam intensity after passage through the experimental apparatus at temperature T, with and without the sample, respectively. The attenuation is produced by the product of the sample mass attenuation coefficient u, the sample density ρ, and the sample length ℓ along the gamma beam. The beam intensity in the absence of a sample, $I_o(T)$, is also a function of temperature and includes temperature-dependent attenuation changes in the chamber walls, crucible, and furnace materials, as well as changes in the gamma beam collimation size. The coefficient u is independent of the physical state of the sample, and therefore is independent of temperature. Eq. (1) can be used to measure the density of a molten material at any temperature where the crucible size $\ell(T)$ is known.

One can show that the densities at two temperatures T_1 and T_2 are related by

$$\rho\,(T_2) = \frac{\rho\,(T_1)}{1 + \alpha\,(T_2 - T_1)} + \frac{\ln\left[I\,(T_1)\,I_o\,(T_2)/I_o\,(T_1)\,I\,(T_2)\right]}{u\ell\,(T_1)\left[1 + \alpha\,(T_2 - T_1)\right]} \quad , \quad (2)$$

which derives from Eq. (1). The inside crucible dimension at T_1 is $\ell\,(T_1)$; the expansion of the crucible is explicitly represented through its linear expansion coefficient α. It is convenient to use the following linear approximation for the beam intensity I_o:

$$\frac{I_o\,(T_2)}{I_o\,(T_1)} = 1 + \varepsilon\,(T_2 - T_1) \quad , \quad (3)$$

where ε is an experimentally determined coefficient which describes the fractional change in I_o with temperature. The error due to this approximation will be discussed below. Eq. (3) can then be substituted into Eq. (2) so that the absolute value of I_o need not be known when its functional temperature dependence has been determined. Eq. (2) is used to determine the density of a molten material at temperature T_2 when the density at T_1 is known.

It has been shown recently that the technique can be used for the measurement of the thermal expansion of isotropic solids.[3] Since the technique yields the product of density and length, $\rho\ell$, an additional relationship is necessary to separate ρ and ℓ in order to deduce $\Delta\ell$, the expansion in sample length along the beam direction. For isotropic materials, the mean volumetric (density) thermal expansion α_ρ can be related to the mean linear thermal expansion α_ℓ by

$$\alpha_\rho \cong -3\alpha_\ell\,(1 - 2\alpha_\ell\,\Delta T) \qquad (4)$$

correct to first order in $\alpha_\ell \Delta T$, where α_ρ and α_ℓ are mean values over a temperature interval $\Delta T \equiv T_2 - T_1$, such that

$$\alpha_\ell = \frac{\ell_2 - \ell_1}{(\Delta T)\,\ell_1} \qquad (5)$$

and

$$\alpha_\rho = \frac{\rho_2 - \rho_1}{(\Delta T)\,\rho_1} \qquad (6)$$

The notation has been simplified by using $\rho_1 = \rho\,(T_1)$, $\ell_1 = \ell\,(T_1)$, etc. It can be shown[1,3] that the expansion coefficient α_ℓ can be found by solution of

$$6x^3 + 3x^2 - 2x - z = 0 \quad , \tag{7}$$

where

$$x \equiv (\Delta T)\alpha_\ell \tag{8}$$

and

$$z = \frac{\ell n\left[I(T_1)I_o(T_2)/I(T_2)I_o(T_1)\right]}{u\rho_1\ell_1} \tag{9}$$

The precision analysis of the gamma attenuation method has been presented elsewhere.[1,2] With typical experimental values, the density ρ (Eq. (2)) and the fractional length expansion x (Eq. 8) can be determined to 0.2%.

The standard form for reporting data on thermal expansion of molten materials is given by

$$\rho(T) = a - b(T - T_m) \tag{10}$$

where a and b are coefficients and T_m is the melting temperature. Since the density $\rho(T_2)$ determined by Eq. (2) depends on the density $\rho(T_1)$ at a fiducial temperature T_1, the error in $\rho(T_2)$ depends strongly on errors in $\rho(T_1)$. As can be seen in Eq. (2) however, the experimental technique yields the relative changes in density with temperature, and not the absolute density itself. We now examine the precision of this technique for determination of the change in density with temperature, represented by b in Eq. (10). First, the function $(\rho_2 - \rho_1)$ is formed from Eq. (2). The partial derivative with respect to each experimental parameter is then formed, and is then weighted by the standard deviation of each parameter to form the rms deviation of the density change, $(\rho_2 - \rho_1)$.[2] Table 1 shows the resultant precision in $(\rho_2 - \rho_1)$ due to the precision in the experimental parameters for a typical set of data. The "worst case" set of data represents a pessimistic esti-mate of uncertainties for each of the parameters; even in this case, the imprecision in the measured density change is small. In particular, we note that rather large percentage standard devia-tions in some of the parameters (ρ, ϵ, and α) produce only small variations in $(\rho_2 - \rho_1)$.

EXPERIMENTAL DETAILS

The gamma attenuation technique involves a determination of the product of the density and length of a material from a measure-ment of the attenuation of a collimated beam of monoenergetic gamma radiation through the material. The gamma detection system has

Table 1. Change in density, $\rho_2 - \rho_1$, calculated from Eq. (2).
"Typical" and "worst" case estimates for the standard
deviations of each parameter are shown. The resultant
rms deviation of $(\rho_2 - \rho_1)$ is calculated as described
in the text. (cps = counts per second)

Parameter	Value	% Standard Deviation	
		Typical	Worst Case
$\rho(T_1)$	7.85 g/cm^3	0.2%	5.0% (7)
u	$0.078 \text{ cm}^2/\text{g}$	0.2	1.0 (2)
$I(T_1)$	21984 cps	0.4*	1.0** (6)
$I(T_2)$	23173 cps	0.4*	1.0** (6)
ε	$2.6 \times 10^{-6}/\text{K}$	10.0	50.0 (5)
T_1	1730 K	1.0	1.5 (4)
T_2	1992 K	1.0	1.5 (3)
$\ell(T_1)$	3.5 cm	0.05	1.0 (2)
α	$9 \times 10^{-6}/\text{K}$	5.0	20.0 (1)
$\rho(T_2) - \rho(T_1)$	-0.208 g/cm^3	1.2%	2.7%

*Represents 5.0% uncertainty in dead time, 0.03% in count rate.
**Represents 10.0% uncertainty in dead time, 0.03% in count rate.
()Shows order ranking of each contribution to the total uncertainty.

been previously described.[2] Briefly, a single channel analysis
counting system is used to detect the beam of 0.662 MeV gamma radiation from a ^{137}Cs source, nominally 4 Ci. The gain of the linear
electronics in the counting system is automatically controlled by a
digital stabilizer, and the temperature of the scintillation detector is thermostatically controlled to within ±0.05°C. These two
features are necessary to achieve count rate stability over extended time periods. All measured count rates are corrected for
the counting system dead time.

The high temperature furnace facility consists of an Astro
(Astro Industries, Inc., Santa Barbara, CA[4]) model 1100 V high
vacuum/controlled atmosphere tungsten mesh furnace and a 50 KVA
power supply. A diagram of the furnace is shown in Fig. 1. The
collimated gamma beam passes through sight ports in the furnace
outer shell and the internal radiation shield pack, and between
the electrodes of the tungsten mesh heater element (7.6 cm diam. x
15.2 cm long). With this furnace, temperatures to 3000 K can be
achieved.

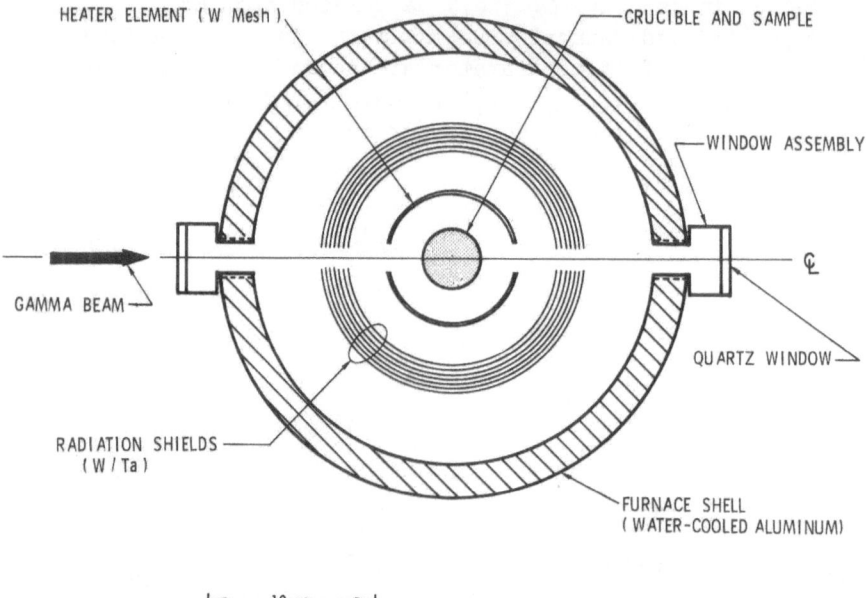

Figure 1.   Schematic diagram of 3000 K high temperature furnace,
horizontal cross section.

Samples were fabricated from VP grade polycrystalline nickel
rod, nominal purity 99.99%, Materials Research Corp., Orangeburg,
NY. For the solid measurements, the sample diameter was nominally
3.4 cm. Prior to solid measurements, the samples were annealed in
an inert atmosphere for over one hour at over 1300 K, and then
furnace cooled for several hours. For both solid and liquid
measurements, the nickel was placed inside right circular cylindrical crucibles (3.8 cm O.D. x 5 cm length) of either MgO or ZrO_2.
A run was also made using an Al_2O_3 crucible of square cross-section,
nominally 2.5 cm on a side. The crucibles were covered to inhibit
nickel vapor loss from the crucible. Post-test examination of the
crucibles indicated no attack by the molten nickel. The mass attenuation coefficient for the sample material was measured at room
temperature. A correction was made for the pathlength in the sample
due to the curvature of the cylinder.[1]

Temperatures were measured using an automatic recording optical pyrometer (0.65 μm) which viewed the crucible side, at the
gamma beam exit, through the furnace sight port. The measured
brightness temperatures were corrected for both the transmission
of the optics and an effective non-unity emittance of the crucible
material inside the heater element.

The effective emittance was determined for each run by calibration to the melting point of nickel (1728 K) as observed by the
gamma counting system. The measured emittances ranged from 0.75
to 0.9, depending on the crucible material. The optical pyrometer
was independently calibrated using blackbody measurements of NBS
melting point standards. The reported temperatures are judged
accurate to 1%.

All measurements were made with a cover gas of high-purity
helium, 3.4×10^4 Pa (gauge). A small flow of He was maintained
into the furnace chamber to prevent fogging of the optical access.

The data reported here include four runs each for solid and
liquid measurements. Data were taken during both heating and
cooling. Temperatures were limited to 2100 K with the MgO crucibles, and 2300 K with ZrO_2 crucibles. Data were taken with the
Al_2O_3 crucible to 1800 K.

The temperature dependence of I_o, the count rate in the
absence of a sample, was measured for each crucible material.
This dependence was adequately described by the linear form in
Eq. (3), and can be predicted by the thermal expansion of the crucible material alone.[1] The largest correction for this temperature dependence is employed for the MgO crucibles, for which $\varepsilon \sim$
4.4×10^{-6}/K.

RESULTS AND DISCUSSION

Solid Nickel

The method described in a previous section was used to deter-
mine the fractional change in length, $\Delta \ell/\ell_{297}$, from the length ℓ_{297}
at ambient (297 K). The data are shown in Fig. 2. The least-
squares quadratic fit to the data is given by

$$\Delta \ell/\ell_{297} = (2.75 \times 10^{-4} \pm 0.047) + (1.27 \pm .11) \times 10^{-3}$$

$$\times (T - 297 \text{ K}) + (4.43 \pm 0.71) \times 10^{-7} (T - 297 \text{ K})^2$$

expressed in percent (rms deviation 0.045). The linear thermal
expansion for nickel to 1500 K as recommended by TPRC[5] is also
shown in the figure. Our results show a larger expansion above
1200 K than is recommended by TPRC; however, their data compila-
tion indicates only a single data set above 1400 K, and none above
1500 K. (Due to this lack of data, they have designated their
values above 1000 K as "provisional.") The results reported here
appear to be the only recent data near the melting point for

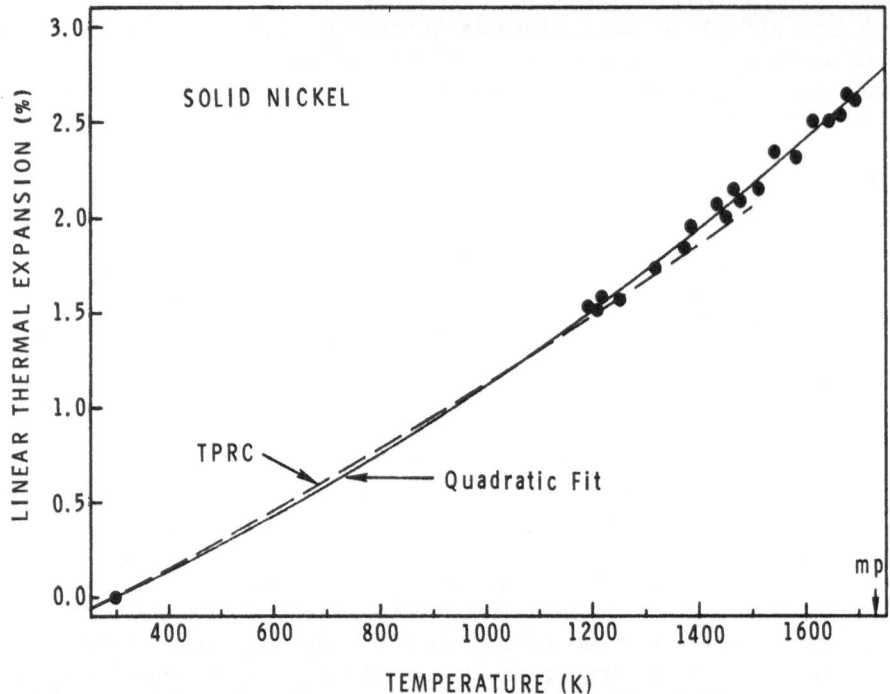

Figure 2.   Linear thermal expansion of solid Ni. Solid curve is
least-squares quadratic fit to the data. Dashed curve
is expansion recommended by the Thermophysical Properties Research Center (TPRC) (provisional above 1000 K).

nickel. The quadratic fit to our data predicts an expansion of
2.72% from ambient to the melt temperature.

Molten Nickel

The data for molten nickel are shown in Fig. 3. A single
measurement was made to determine $\rho(T_1)$ slightly above the melting
point (mp 1728 K). This value agreed with the recommended melting
point liquid density by Crawley[6] within an experimental error of
0.2%. All other data points were calculated from Eq. (2) using
this value. The linear least-squares fit to the data is given by
$\rho(T) = (7.850 \pm 0.003)$ g/cm^3 - $(7.35 \pm 0.09) \times 10^{-4} \times (T - 1728$ K$)$
g/cm^3 K (rms deviation 0.008 g/cm^3). Also shown on the figure is
the thermal expansion of molten nickel recommended in the review
article by Crawley,[6] expressed as $\rho(T) = 7.89 - 9.91 \times 10^{-4} (T - 1728$ K$)$, in g/cm^3. Our data suggest an expansion coefficient significantly smaller than the value recommended by Crawley. A com-
parison of much of the recent available data for molten nickel is
given in Table 2, which shows the significant spread in data
reported in the literature.

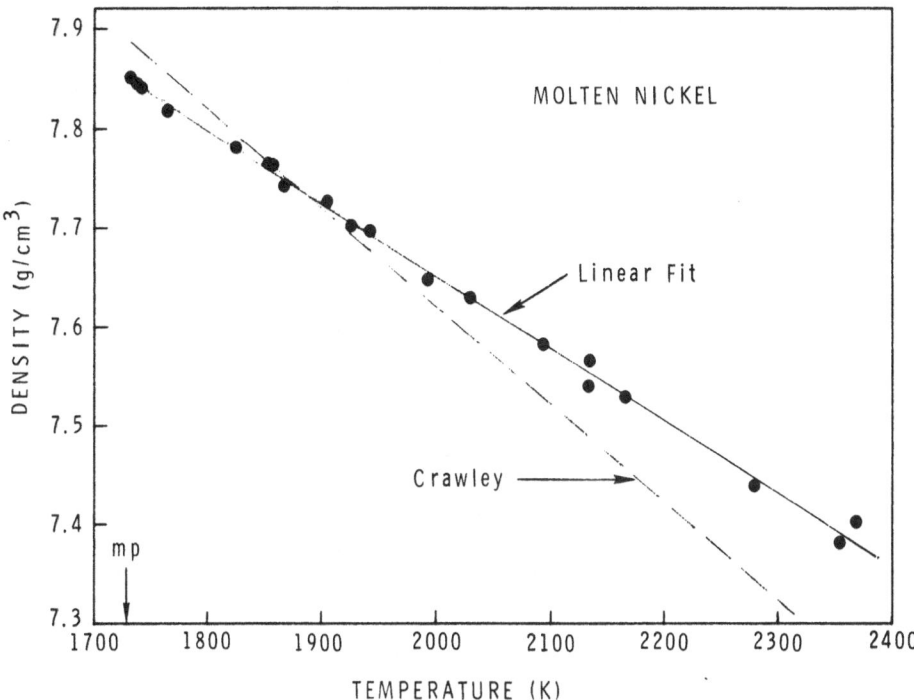

Figure 3. Density of molten nickel vs temperature. Solid curve is
           least-squares linear fit to the data. Dashed curve is
           linear fit recommended by Crawley (see text).

Table 2. Thermal Expansion of Molten Nickel
          from Several Investigations

Reference	Year	Method	$b, 10^{-4} \ g/cm^3 \ K$
Saito et al.[7]	1969	Maximum Bubble Pressure (MBP)	12.1
Lucas[8]	1961	Maximum Bubble Pressure	11.9
Kirshenbaum and Cahill[9]	1963	Direct Archimedean	11.5
Saito and Sakuma[10]	1970	Levitated Drop	10.8
Shiraishi and Ward[11]	1964	Levitated Drop	10.2
Saito et al.[12]	1969	Levitated Drop & MBP	9.9
Popel et al.[13]	1969	Large Drop	8.7
This work	1979	Gamma Attenuation	7.35 ± 0.09
Vertman et al.[14]	1964	Large Drop	6.4

    The spread in data shown in Table 2 suggests large systematic
differences between the various measurement techniques. The current work represents the only data on nickel using the gamma
attenuation technique. It is interesting to note the correlation
between measurement method and magnitude of thermal expansion. A
definitive comparison of the techniques and the systematic errors
associated with each would be quite helpful in this area.

    As for our data, the uncertainty in the fit parameter b is on
the order of the expected precision in the change in density due
to propagation of random errors as discussed in a previous section.
We have considered a number of possible sources of systematic
error with this technique. The combined effects of cylindrical
samples, gamma beam diameter uncertainty, and offset of the beam
from the sample center amount to less than 1% uncertainty in the
value of $\ell(T_1)$. Data from heating and cooling, as well as post-
test examination, indicate no change in crucible inside diameter
due to chemical attack. The measurement of the temperature dependence of I_o accounts for temperature-induced changes in collimation, window materials, inert gas atmosphere, and crucible attenuation. Since the mass attenuation coefficient was measured rather
than inferred from handbook values, it includes geometrical
effects due to gamma scattering and non-ideal collimation. (We
note that our measurements of the attenuation coefficient are
within a few percent of published literature values.) No difference in gamma attenuation at high temperatures was observed
between empty crucibles and crucibles filled with nickel vapor.
This measurement removes any question of excessive attenuation due
to nickel vapor species outside the crucible. Measurements on
molten tin and lead to 1300 K using a lower temperature furnace
facility with the same measurement technique agreed very well with
data obtained from other techniques.[15] We have extended the
measurements on molten Pb to 1500 K using the apparatus described
in this paper, and have found excellent agreement with our previous
results to 1300 K. To account for the differences in thermal expansion solely on the basis of temperature measurement error would
require errors in excess of 100 K at 2000 K. For our data, a
melting point calibration of the pyrometer measurement system for
each run makes such temperature errors extremely unlikely. As
shown in the precision analysis section, large uncertainties in
melting point density, crucible expansion coefficient α, and ε
(I_o temperature dependence) are not sufficient to explain the dis-
crepancies with other investigations. In summary, we have been
unable to identify any significant source of systematic experimental error in the gamma attenuation technique, as utilized in
these experiments.

    An interesting correlation between the thermal expansion co-
efficient of liquid metals and their boiling and melting temperatures has been observed by Steinberg.[16] He notes that previous

measurements of the thermal expansion coefficients for Fe, Co, and Ni are significantly higher than predicted by his correlation. However, the data reported here agree with this correlation quite well; for nickel, the correlation predicts $b \sim 7 \times 10^{-4}$ g/cm^3 K. We are planning measurements on iron and cobalt to investigate any similar behavior.

CONCLUSIONS

The linear thermal expansion of solid nickel was measured from ambient to the melting point. Above 1200 K, the expansion was observed to be slightly larger than expected from provisional values given by TPRC. The thermal expansion of molten nickel was measured from the melt to 2300 K. The expansion coefficient was found to be significantly smaller than most previous measurements using other techniques. A precision analysis of random experimental errors and a search for systematic errors have been unable to identify any source to explain differences between our results and previous measurements. Our data for the thermal expansion of molten nickel agree with the Steinberg correlation for the thermal expansion of liquid metals much better than most previous data.

REFERENCES

1. W. D. Drotning, "Thermal Expansion and Density Measurements of Molten and Solid Materials at High Temperatures by the Gamma Attenuation Technique," SAND 79-0074, Sandia Laboratories, Albuquerque, NM (January, 1979). Available from: National Technical Information Service (NTIS), U.S. Dept. of Commerce, 5285 Port Royal Road, Springfield, VA 22161.

2. W. D. Drotning, in Thermal Expansion 6 edited by I. A. Peggs (Plenum, NY, 1978), pp. 83-96.

3. W. D. Drotning, Rev. Sci. Instrum. 50, 1567 (1979).

4. Reference to a company product implies neither a recommendation by Sandia Laboratories or the U.S. Department of Energy nor a lack of suitable substitutes.

5. Y. S. Touloukian et al. (Ed.), "Thermal Expansion," in Thermophysical Properties of Matter, Vol. 12 (IFI-Plenum, New York, 1977).

6. A. F. Crawley, Inter. Metall. Rev. 19, 32-48 (1974).

7. T. Saito, M. Amatatu and S. Watanabe, Tohoku Daigaku Senko Seiren Kenkyusho 25, 67 (1969).

8. L. D. Lucas, Mem. Sci. Rev. Met. 61, 1 (1964); also "Liquid Density Measurements," Chap. 7B, in Techniques of Metals Research Vol. IV, Part 2, (edit. R. A. Rapp), Interscience, New York, NY, 1970.

9. A. D. Kirshenbaum and J. A. Cahill, Trans. Amer. Soc. Metals 56, 281 (1963).

10. T. Saito and Y. Sakuma, Sci. Rep. Research Inst. Tohuko Univ. 22, [A], 57 (1970).

11. S. Y. Shiraishi and R. G. Ward, Can. Met. Quart. $\underline{3}$, 117
     (1964).
12. T. Saito, Y. Shiraishi and Y. Sakuma, Trans. Iron Steel Inst.
     Japan $\underline{9}$, 118 (1969).
13. S. I. Popel, L. M. Shergin and B. V. Tsarevskii, Zhur. Fiz.
     Khim. $\underline{43}$, 2365 (1969).
14. A. A. Vertman, A. M. Samarin and E. S. Filippov, Dokl. Akad.
     Nauk SSSR $\underline{155}$, 323 (1964).
15. W. D. Drotning, High Temp. Sci. $\underline{11}$, 265 (1979).
16. D. J. Steinberg, Metall. Trans. $\underline{5}$, 1341 (1974).

THERMAL EXPANSION OF FeF$_2$ CRYSTALS

A.S. Pavlovic

Department of Physics
West Virginia University
Morgantown, West Virginia 26506

INTRODUCTION

Many experimental and theoretical studies have been made of the properties of ferrous fluoride (FeF$_2$).[1-11] However, no detailed measurements of the thermal expansion of FeF$_2$ below room temperature have been reported. This paper presents the results of such an investigation of the thermal expansion of FeF$_2$ with particular emphasis on its behavior in the vicinity of the Neel temperature.

Iron difluoride is a compound which has a crystal structure that is isomorphic with the difluorides of manganese, nickel and cobalt. It has a tetragonal, rutile structure with a c/a ratio of 0.703.[1] The Fe^{2+} ions are located on the corner and body-centered sites of the unit cell. From room temperature to about 78° K, it is paramagnetic. At 78 K, it undergoes a second order phase transition to a tetragonal antiferromagnetic phase where the magnetic moments of the Fe^{2+} ions are aligned along the c-axis; the corner ions are anti-parallel with the body-centered ions.[4] The magnetic properties of FeF$_2$ are characterized by a large anisotropy throughout the paramagnetic and anti-ferromagnetic domains. This anisotropy has been shown to arise mainly from the interaction of a single-ion with its crystalline field.[2,5,6,9]

The measurements of the thermal expansion of FeF$_2$ in the a- and c-directions given below exhibit a large anisotropy. At the Neel temperature, the coefficient of thermal expansion in the c-direction has a λ-type anomaly just like the specific heat. An estimate of the pressure dependence of the Neel temperature of $dT_N/dp = 0.36$ deg.Kbar^{-1} has been calculated. This is in fair agreement with the experimentally determined value of 0.27 deg.Kbar^{-1}.

EXPERIMENTAL PROCEDURE

The samples employed in this study were cut from a large single crystal of optical quality kindly provided by Dr. R. Fiegelson of Stanford University. These had the rough dimensions of 5mm x 5mm x 3mm. The orientations of the sample were determined by Laue back reflection of x-ray diffraction patterns which were within ± 0.5° of the exact orientations.

Strain gages in a d.c. wheatstone bridge circuit were used to measure the thermal expansion. The data was corrected for gage factor temperature variation and the thermal expansion of the high purity copper dummy. The data points are the average of measurements made on four crystals with two different gages on each. The spread of the results was about 4%.

The data was taken at temperatures which were stabilized to ± 0.05 K by means of an Artronix controller and a copper-constantan thermocouple. Sample temperatures were measured using a calibrated copper-constantan thermocouple and a Leeds and Northrup K-5 photentiometer. In the vicinity of the transition temperature, the data was taken at 0.1 K intervals while outside this region 5 to 10 K intervals were employed.

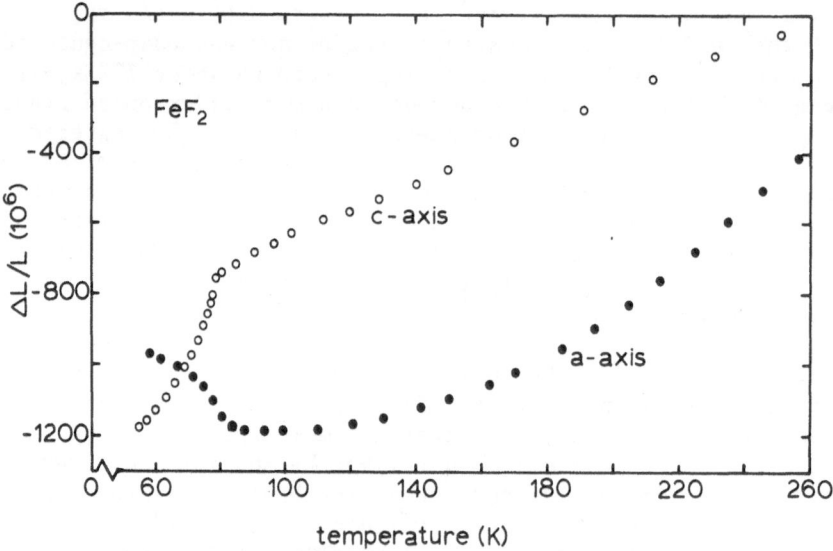

Fig. 1.   The thermal expansion of FeF$_2$ along the a- and c-directions as a function of temperature.

RESULTS AND DISCUSSION

     The thermal expansion measurements presented here were made
with respect to the length of the sample at 293 K. Figure 1 shows
the thermal expansion of FeF$_2$ along the a- and c-directions. Both
exhibit a contraction as the temperature decreases from room temperature. A large anisotropy develops in the thermal expansion with
the crystal constracting more rapidly in the a-direction than in the
c-direction. At about 80 K, the a-direction thermal expansion levels
off and begins to expand below 80 K. The c-direction thermal expansion begins to contract at a faster rate at and below 80 K. Inflection points occur for both curves at T_N = (78.2 ± 0.1)K. The data
from room temperature to about 90 K has been fitted to a power expansion in the temperature up to the third power by a least squares
computer program. The expression for the thermal expansion in the
a-direction is

$$(\Delta \ell/\ell)_a = -9.02 \text{x} 10^{-4} - 5.79 \text{x} 10^{-6} T + 2.96 \text{x} 10^{-8} T^2 + 2.31 \text{x} 10^{-12} T^3$$

and in the c-direction

$$(\Delta \ell/\ell)_c = -1.57 \text{x} 10^{-3} + 6.33 \text{x} 10^{-6} T - 2.81 \text{x} 10^{-9} T^2 - 1.20 \text{x} 10^{-12} T^3.$$

     The coefficients of linear thermal expansion were determined
by graphical methods. Figure 2 exhibits the coefficients as a function of temperature. In the a-direction the coefficient of linear

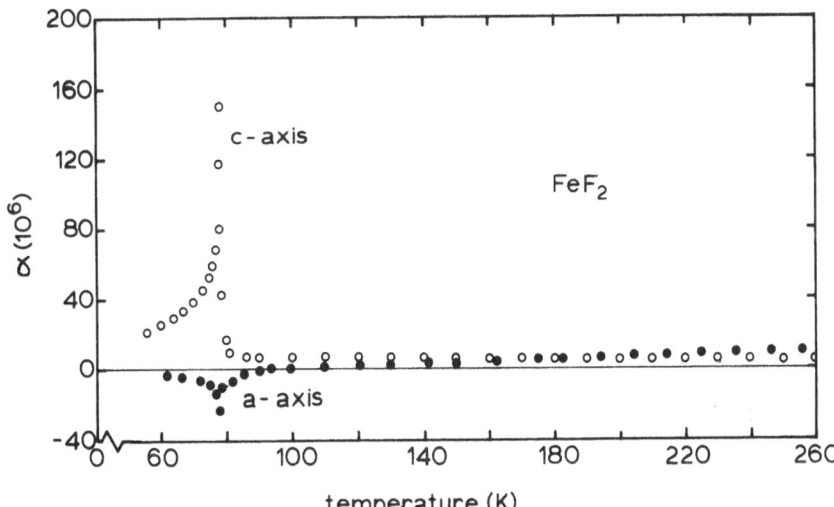

Fig. 2.   The coefficient of linear thermal expansion of FeF$_2$ in
            the a- and c-directions as a function of temperature.

thermal expansion decreases continuously and uniformly from a positive value at room temperature until it becomes negative at about
95 K. Below 95 K, it has a small but sharp negative peak at 78.2 K.
The coefficient of linear thermal expansion in the c-direction is
almost constant from room temperature to about 90 K where it begins
to rise slowly; then at 78.2 K, it takes a sudden jump to a large
value of 152×10^{-6}. Below 78.2 K it drops rapidly at first then
slowly approaches zero. This λ-type anomaly is similar to the specific heat anomaly reported by Stout and Catalano.[3] They pointed out
that FeF_2 had the steepest and largest λ-anomaly of all the difluorides. We have determined α_c for NiF_2 and MnF_2 and have found
that FeF_2 has the sharpest and largest rise of the three. The
abruptness of the antiferromagnetic transition is even more apparent if the coefficient of linear volume expansion is calculated
as a function of temperature. This is shown in Figure 3. Notice that
it goes negative just before the occurrence of the transition,i.e.,
at 85 K and then very suddenly rises.

It was not possible to determine the anisotropic Gruneisen
parameters because the elastic compliances for FeF_2 are not available. Nevertheless, the coefficient of linear volume expansion was
plotted against the specific heat at constant pressure as shown

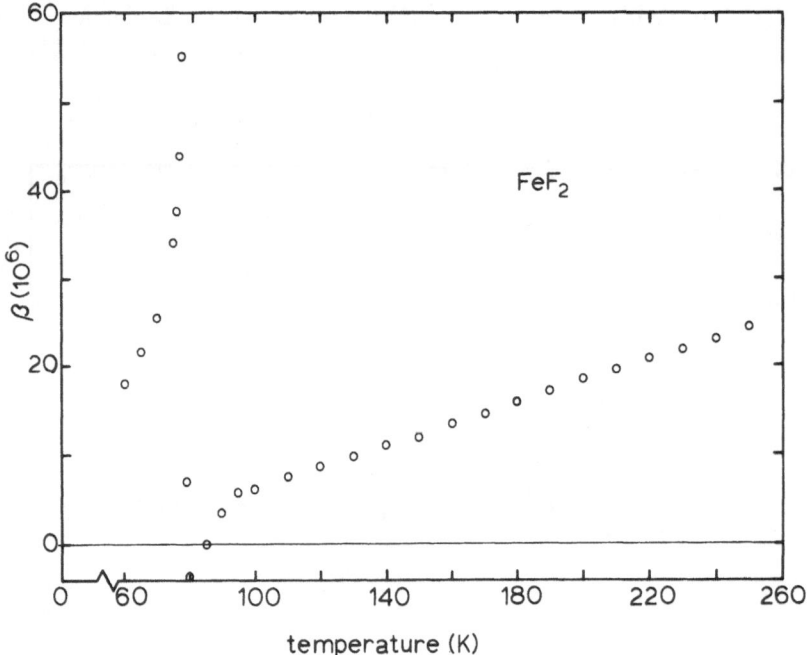

Fig. 3.   The temperature dependence of the coefficient of linear
          volume expansion of FeF_2.

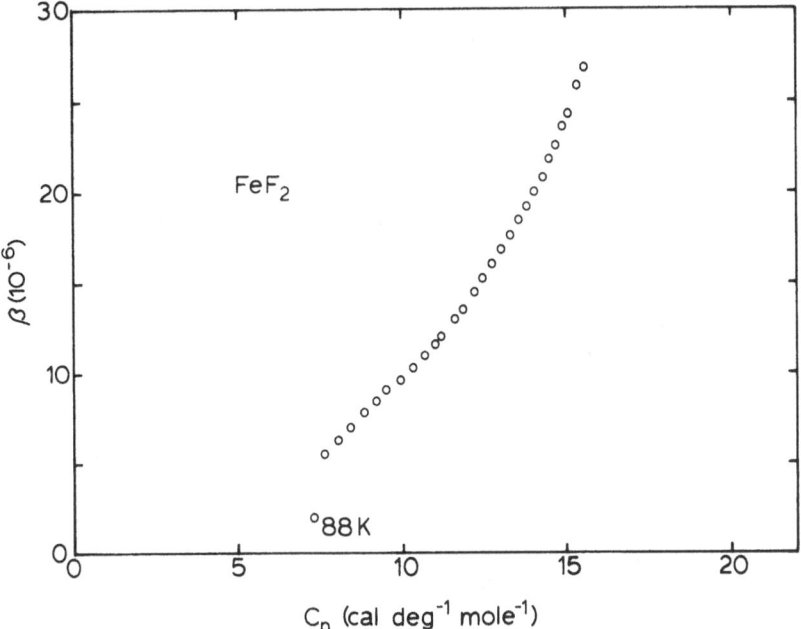

Fig. 4.   A plot of the coefficient of linear volume expansion versus the specific heat at constant temperature between
room temperature and 88 K.

in Figure 4. For solids which have tetragonal symmetry, the coefficient of linear volume expansion (β) can be written as

$$\beta = C_p[2(S_{11} + S_{12} + S_{13})\gamma_1 + (2S_{13} + S_{33})\gamma_3]/V$$

where C_p is the specific heat at constant temperature, S_{ij} are the
elastic compliances, γ_k are the components of the Gruneisen tenor,
and V is the volume.[12] Since the β versus C_p curve in Figure 4 is
not a straight line, the coefficient of C_p is not a constant.

The λ-anomaly at the Neel temperature indicates that the antiferromagnetic transition is a phase transition of second order.
Therefore, the pressure coefficient of the transition temperature,
dT_N/dp, can be estimated by the Ehrenfest relation.

$$dT_N/dp = \frac{T_N V \Delta\beta}{\Delta C_p}$$

where p is the hydrostatic pressure, V is the volume at T_N and ΔC_p
is the change in the specific heat at T_N and $\Delta\beta$ is the change in
the coefficient of linear volume expansion at T_N. In the case of

FeF_2, $T_N = 78.2$ K, $V = 21.62 \times 10^{-6} m^3/mole$, $\Delta C_p = 5.34$ cal·deg^{-1} mole^{-1} and $\Delta\beta = 48$ deg^{-1} which when substituted in the Ehrenfest relations gives $dT_N/dp = 0.36$ deg Kbar^{-1}. Garcia and Ingalls have made a Mössbauer measurement of the pressure dependence of the Neel temperature and found that $dT_N/dp = 0.27 \pm 0.03$ deg·Kbar^{-1}. The estimate of dT_N/dp is in fair agreement with the experimental result.

ACKNOWLEDGMENT

This work was partially supported by the NSF. The author gratefully acknowledges the assistance of Mr. Jason Cook and the generosity of Dr. R. Fiegelson in providing the FeF_2 crystal.

REFERENCES

1.   J.W. Stout and S.A. Reed, The crystal structure of MnF_2,FeF_2, CoF_2, NiF_2 and ZnF_2, J.Am. Chem. Soc. 76:5279 (1954).

2.   J.W. Stout and L.M. Matarrese, Magnetic anisotropy of the iron-group fluorides, Rev. Mod. Phys. 25:338 (1953); S. Foner, High field magnetic moment and antiferromagnetic resonance measurements in $\alpha-Fe_2O_3$, CoF_2, FeF_2 and $(MnF_2)_{1-x}(ZnF_2)_x$, in Proceedings of the International Conference on Magnetism, Nottingham, 1964 (The Institute of Physics and Physical Society, London, 1965).

3.   J.W. Stout and E. Catalano, Heat capacity and entropy of FeF_2 and CoF_2 from 11 to 300° K. Thermal anomalies associated with antiferromagnetic ordering, J. Chem. Phys. 23: 1803 (1955).

4.   R.A. Erikson, Neutron diffraction studies of antiferromagnetism in manganous fluoride and some isomorphous compounds, Phys. Rev. 90:779 (1953).

5.   R.C. Ohlmannn and M. Tinkham, Antiferromagnetic resonance in FeF_2 at far-infrared frequencies, Phys. Rev. 123:425 (1961).

6.   V. Jaccarino, Nuclear resonance in antiferromagnets in magnetism, G.T. Rado and H. Suhl, eds., Academic Press, Inc., New York (1965), Vol. 2A.

7.   Y. Shapira, Paramagnetic-to-antiferromagnetic phase boundaries of FeF_2 from ultrasonic measurements, Phys. Rev. B 2:2725 (1970).

8.   D.P. Johnson and R. Ingalls, Mossbauer studies of lattice dynamics, fine and hyperfine structure of divalent Fe^{57} in FeF_2, Phys. Rev. B 1:1013 (1970).

9.   B.R. Cooper, Sublattice magnetization and resonance frequency of antiferromagnets with large uniaxial anisotropy, Phys. Rev. 120: 1171 (1960).

10. M.E. Lines, Sensitivity of Curie temperature to crystal-field anisotropy, II. FeF_2, Phys. Rev. 156:543 (1967).

11. T. Tanaka, L. Libelo and R. Kligman, Application of the
      cluster-variation method to the uniaxial antiferromagnet
      FeF$_2$, Phys. Rev. 171:531 (1968).
12. Thermophysical Properties of Matter, Vol. 13, Thermal Expansion
      Nonmetallic Solids, p. 9a, Y.S. Touloukian, ed., Plenum,
      New York, 1977.
13. G.A. Garcia and R. Ingalls, The pressure dependence of the Neel
      temperature in FeF$_2$, J. Phys. Chem. Solids 37:211 (1976).

EXFOLIATION OF GRAPHITE*

D. D. L. Chung

Department of Metallurgy and Materials Science and
Department of Electrical Engineering
Carnegie-Mellon University
Pittsburgh, PA 15213

INTRODUCTION

When a graphite intercalation compound [1] is heated past a critical temperature, a large expansion along the c-direction occurs, giving the compound a puffed-up appearance. This phenomenom is known as exfoliation. Ubbelohde [2] observed that graphite-Br_2 exfoliated at 350°C from 3 mm to approximately 35 mm. Exfoliation has also been observed in graphite-$FeCl_3$ [3], graphite-$AlCl_3$ [4], and graphite intercalated with a mixture of HNO_3 and H_2SO_4 [4]. The exfoliation of graphite-$FeCl_3$ has been used to manufacture Grafoil [5]; the exfoliation of graphite-(HNO_3+H_2SO_4) has been used for making a thermal insulator for molten metals [6]. In spite of the numerous practical applications of exfoliation, relatively little work has been done to understand and characterize this unusual phenomenon.

In this work, thermomechanical analysis (TMA) was used to obtain quantitatively measurement of the expansion; scanning electron microscopy (SEM) was used to charaterize the microstructural changes which accompany exfoliation; effluent gas detection and thermogravimetry were used to investigate the effect of exfoliation on the intercalate concentration in the compound.

---

*Research sponsored by the Air Force Office of Scientific Research, Air Force Systems Command, USAF, under Grant No. AFOSR-78-3536. The United States Government is authorized to reproduce and distribute reprints for Governmental purposes notwithstanding any copyright notation hereon.

The large expansion that characterizes exfoliation is itself a form of thermal instability; which results in significant changes in the electrical and mechanical properties along the c-direction, although this same phenomenon can be advantageously used in practical applications [5,6]. In addition to the instability in sample thickness is the instability in intercalate concentration, as observed by gas detection and thermal gravimetry [7].

The intercalate species used in this work was bromine, which is the most widely studied acceptor intercalate species. Graphite-bromine was prepared by exposure of pristine graphite to bromine vapor in equilibrium with bromine liquid at room temperature. This resulted in $C_{16}Br_2$, the saturated compound having a stage 2 structure. Removal of the compound from the bromine vapor led to intercalate desorption, which resulted in more dilute compounds.

EXPERIMENTAL

Thermomechanical Analysis

Thermomechanical analysis (TMA) was performed by using a Perkin-Elmer Model TMS-2 thermomechanical analyzer, in which a linear variable differential transformer (LVDT) monitored the position of the probe which made contact with the sample.

In TMA, one measures the expansion versus the temperature, as shown in Fig. 1, which was obtained during successive exfoliation and collapse cycles of dilute graphite-Br_2. Of interest is the mechanical hysterisis observed on cooling. Very little contraction was observed until fairly low temperatures were reached. At low temperatures, however, contraction occurred rapidly and fairly completely. The second exfoliation (indicated by dashed lines) occurred at temperatures lower than those of the first exfoliation (indicated by solid lines), and was essentially completed within a much smaller temperature range than that needed for the first exfoliation. Furthermore, the expansion curve for the second exfoliation is smooth and steep, in contrast to the shoulders observed in the first exfoliation curve. Exfoliation-collapse cycles subsequent to the second exhibit the same TMA results as the second cycle.

Exfoliation has been found to be reversible for dilute compounds, as illustrated in Fig. 1. However, for concentrated compounds, the large and violent expansion during exfoliation was accompanied by considerable permanent deformation, so that the compound remained expanded on subsequent cooling. Thus, the reversibility of exfoliation depends on the intercalate concentration. The reversibility for dilute compounds promises numerous practical applications of the exfoliation phenomenon.

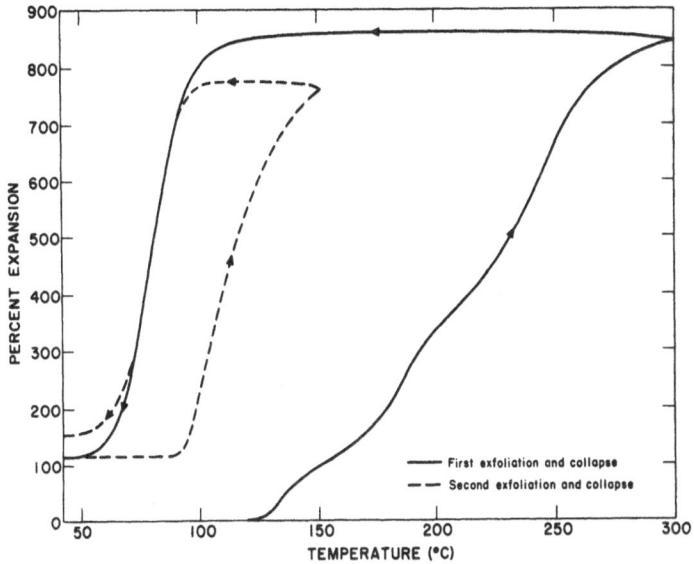

Fig. 1   Expansion vs. temperature curves obtained during
         successive exfoliation and collapse cycles of
         dilute graphite-Br$_2$.

Scanning Electron Microscopy

Exfoliation can be easily observed by the naked eye. However,
in order to investigate the microstructural effects of exfoliation,
we have used scanning electron microscopy (SEM). The microscopic
examination was performed by using a JEOL JMS 35 scanning electron
microscope, which is equipped with an energy dispersive x-ray analyzer for elemental chemical analysis.

Figure 2 shows SEM photographs of the a-faces of exfoliated
graphite-Br$_2$ based on highly-oriented pyrolytic graphite (HOPG),
grafoil and natural graphite flakes. These samples were produced
by heating graphite-bromine in N$_2$ gas at 1 atm pressure to ~200°C.
In all samples, numerous cracks perpendicular to the c-direction
were observed. However, the amount of expansion during exfoliation
is smaller for graphite-Br$_2$ based on natural graphite flakes than
for graphite-Br$_2$ based on HOPG or grafoil. This suggests that a
large c-axis thickness of the sample is favorable for more extensive exfoliation.

Although not shown in Fig. 2, negligible exfoliation was produced in graphite-Br$_2$ based on isotropic graphite (UCAR Grade ATJS)
and graphite fibers (PAN-based), even if the samples were heated
to ~800°C. This observation indicates that the alignment of the
c-axes of different grains is necessary for exfoliation.

HIGHLY-ORIENTED
PYROLYTIC GRAPHITE

GRAFOIL

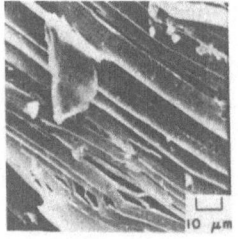

NATURAL GRAPHITE
FLAKE

Fig. 2   SEM photographs of the a-faces of exfoliated
graphite-Br_2 based on different pristine
graphite materials.

Effluent Gas Detection

The detection of desorbed intercalate from a sample upon heating was achieved by using an effluent gas analyzer incorporated in
a differential scanning calorimeter (Perkin-Elmer DSC-1B). The
analyzer consisted of a two-thermister bridge circuit which monitored the thermal conductivity of the DSC sample holder purge gas
relative to the gas which bypassed the sample holder. The system
was purged at 30 cc/min with dry nitrogen. A weighed graphite-Br_2
sample was placed in a platinum pan and mounted in the DSC sample
holder which was purged for about 10 minutes at room temperature.
The sample temperature was increased at either 5 or 10°C/min. A
Columbia Scientific Industries integrator was used to record the
effluent analyzer output; the acquisition rate was 5 or 10 per
minute.

Figure 3 shows the effluent gas analyzer response for a dilute
graphite-Br_2 sample. The sample was originally a saturated lamel-
lar compound (83 wt. %Br_2); prior to this temperature scan, it had

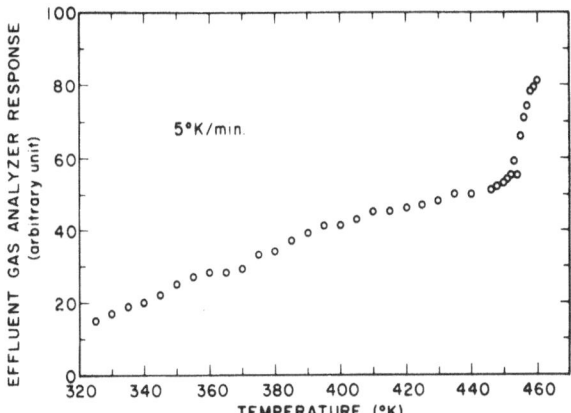

Fig. 3   Effluent gas detector response as a function of
          temperature, showing desorption associated with
          exfoliation near 177°C. The sample contained
          21.5 wt. %Br$_2$ before heating.

been desorbed at room temperature and by heating to about 117°C,
and it contained 21.5 wt. %Br$_2$. The temperature of the sample was
scanned from room temperature to 187°C at 5°C/min. In Fig. 3, the
analyzer trace increased slowly (probably due to baseline drift at
this high sensitivity) until about 177°C were the amount of desorbed intercalate increased sharply. This increase is attributed
to extensive exfoliation, which was visually confirmed at 176°K
for this sample.

Thermogravimetry

     The thermal gravimetric measurement was performed by using a
Perkin-Elmer electronic microbalance (Autobalance Model AD-2Z),
which has a maximum sensitivity of 0.1 μg. The sample was placed
on a Pyrex pan which was suspended by a Pyrex hangwire. A 19 mm
I.D. Pyrex tubing enclosed the hangwire and the sample pan. During the measurement, the tube was slowly purged with argon at
approximately 20 cc/min. A low mass furnace surrounded the sample
pan and was controlled by a Theall Engineering Model TP-2000 temperature programmer. The sample temperature was measured by plac-
ing a chromel-alumel thermocouple immediately below (within 2 mm)
the sample pan. All temperature scans were performed at a heating
rate of 2°C/min.

     Figure 4 is a plot of intercalate concentration versus temper-
ature for three samples which were obtained by desorbing saturated
graphite-bromine compounds at various temperatures to the starting
concentration, which was less than 30 wt. %Br$_2$ in order to avoid
the very high desorption rate at high temperatures for high concentration samples. The samples were heated at 2°C/min, and held
isothermally at 120, 130, or 140°C for 50 minutes before being
heated to higher temperatures. The sample associated with curve 1

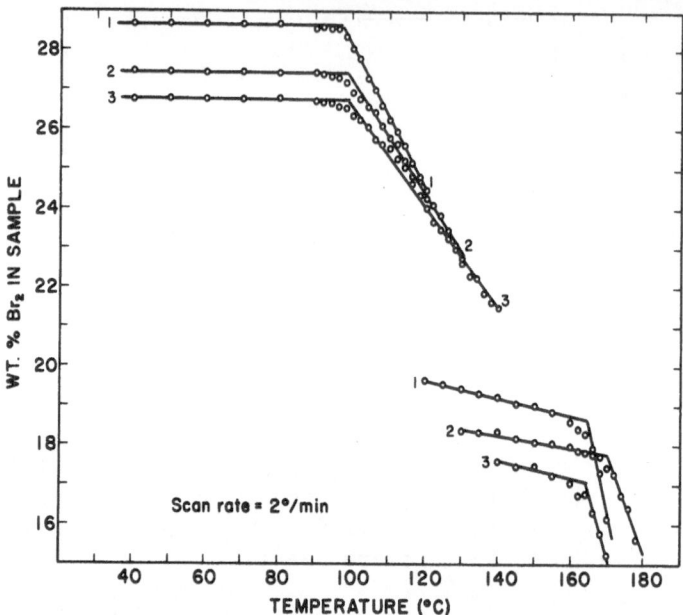

Fig. 4   Thermogravimetric analysis (TGA) experimental
          curves taken at 2°C/min for three relatively
          dilute graphite-Br$_2$ samples. The scan was inter-
          repeated for 50 minutes at 120°C for Sample No. 1,
          at 130°C for Sample No. 2, and at 140°C for
          Sample No. 3.

was initially desorbed at 60°C and was held at 120°C. The samples
yielding curves 2 and 3 were desorbed at 50 and 70°C, respectively,
and were held at 130 and 140°C, respectively.

    As shown in Fig. 4, the sample weight decreases sharply at
two distinct temperatures. The first decrease occurs at approximately 98°C and corresponds closely to the temperature associated
with the intralayer intercalate position order-disorder transformation [11,12]. The second decrease occurs at ∿165-170°C and is
associated with exfoliation.

DISCUSSION

    The extensive expansion that characterizes exfoliation was
found to be accompanied by an increase in the rate of intercalate
desorption, as observed by effluent gas detection and thermogravimetry. However, this does not imply that exfoliation necessarily
involves desorption because the amount of desorption during the
reversible exfoliation of very dilute compounds is negligible. It
is believed that reversible exfoliation is due to the expansion on
heating and contraction on cooling of gas cells [8,9]. At relatively high intercalate concentrations, some gas cells rupture,
causing intercalate desorption. Widespread rupture and desorption

lead to permanent deformation, as observed for the exfoliation of concentrated compounds.

The alignment of the c-axes of different grains minimizes the interference among the grains when each grain elongates along its c-axis [10]. Without such an alignment, the stress exerted by one grain on another can be sufficient to make exfoliation impossible, as we have found by observing the behavior of compounds based on isotropic graphite and graphite fibers. Of interest is the comparison of the exfoliation behavior of pyrolytic graphite and highly-oriented pyrolytic graphite (HOPG). By using pyrolytic graphite, Martin and Brocklehurst [8] observed a maximum expansion of 380%; by using HOPG, we observed a maximum expansion of over 850%. This comparison indicates that the tendency toward exfoliation increases with increasing perfection of aligment of the grains along the c-direction.

A large thickness of the sample along the c-direction was also found to favor extensive exfoliation. This behavior is related to the observation that a large basal plane stack height (L_c) favors exfoliation [11]. It is probably due to the availability of large gas cells in thick samples.

A large difference in exfoliation temperature between first and second exfoliation cycles was observed by TMA. This observation is consistent with previous TMA work on pyroloytic graphite [8] and differential thermal analysis results [12]. It suggests that first exfoliation involves the unpinning of certain layers linked by defects. Once this unpinning has been achieved, subsequent exfoliation requires less energy.

Another difference between first and second exfoliation cycles is that the TMA curve for the first exfoliation cycle exhibits shoulders which are reproducible, as can be seen in Fig. 1, but the TMA curve for the second exfoliation exhibits no shoulders. The origin of the shoulders is presently not clear.

For a complete understanding of exfoliation, it is clear that much more experimental investigation is necessary to characterize both the exfoliation process and the exfoliated compound. This paper just represents a new step in this direction.

REFERENCES

1. Proceedings of the First International Conference on the Intercalation Compounds of Graphite, 1977, Mater. Sci. Eng., 31 (1977).

2. A. R. Ubbelohde, Brit. Coal Util. Res. Assoc. Gaz. 51, 1 (1964).

3. R. E. Stevens, S. Ross, and S. P. Wesson, Carbon <u>11</u>, 525 (1973).

4. M. B. Dowell, Ext. Abstr. Program--Bienn. Conf. Carbon <u>12</u>, 35 (1975).

5. Union Carbide Trademark, U.S. Patent No. 3,404,061.

6. H. Mikami, Japan. Kokai 76 96,793 (1976).

7. J. S. Culik and D. D. L. Chung, Mater. Sci. Eng., to be published.

8. W. H. Martin and J. E. Brocklehurst, Carbon 1, 133 (1964).

9. S. H. Anderson, J. S. Culik, and D. D. L. Chung, Ext. Abstr. Program.--Bienn. Conf. Carbon, <u>14</u>, 262 (1979).

10. A. L. Sutton and V. C. Howard, J. Nuclear Materials 7, (1)58 (1962).

11. M. B. Dowell, Ext. Abstr. Program--Bienn Conf. Carbon 12, 31 (1975).

12. C. Mazieres, G. Colin, J. Jegoudez, and R. Setton, Carbon <u>13</u>, 289 (1975).

THERMAL-EXPANSION STUDIES OF HIGH-DAMPING Mn-Cu ALLOYS

D. K. Chatterjee* and P. Venkateswararao

Defense Metallurgical Research Laboratory
Hyderabad-500258
India

ABSTRACT

The thermal expansion of Mn-Cu alloys was measured in order to study the phase transformation and to assess the stability of the phases responsible for the high damping capacity observed in these alloys. High internal stresses build up during these transformations, and a large volume contraction could provide the necessary mechanism for absorption of vibrational energy.

INTRODUCTION

Solid solutions of manganese and copper have a number of unique features.[1,2] The most characteristic feature of the Mn-Cu system is that equilibrium is attained at a very slow rate, the time required for reaching equilibrium varying in an unsystematic manner with composition. The temperature coefficient of resistance in Mn-Cu alloys exhibits unusual behavior, first decreasing gradually with decreasing Cu content up to about 50% Cu and then increasing rapidly below 35% Cu. This coefficient is negative for alloys containing 80-35% Cu, and the trend persists up to high temperatures. Alloys in this compositional range also exhibit high vibration-damping capacity, the magnitude varying with composition.

The negative temperature coefficient of resistance and very high damping capacity suggest that phase transformation occurring in this alloy could be responsible for this behavior. This paper

*Presently at Systems Research Laboratories, Inc., 2800 Indian Ripple Road, Dayton, OH 45440.

45

presents results obtained from dilatometric investigations on the
Mn-Cu alloy system. The experimental results suggest a possible
hypothesis to explain the high damping characteristic of these
alloys.

EXPERIMENTS AND RESULTS

Alloys containing 15, 30, 35, 40, 45, and 55 wt% Cu were prepared for this investigation. Details of the alloy melting and
homogenization are given elsewhere.[3] Properly homogenized alloys
were quenched from 800°C. Cylindrical specimens 20 mm long and
3.5 mm in diam. were prepared, and a dilatometric study was made
in a Leitz's Universal Dilatometer at a heating rate of 5°C/min.
under 2×10^{-2} Torr vacuum. Dilatometric recordings were obtained
by differential methods using a Chronin standard. Experimental
dilatometric curves are shown in Fig. 1. Transformations were
found to occur in these alloys roughly over three temperature
ranges: 300-400°C, 400-500°C, and 500-600°C. Inspection of the
curves shows that the rapidity of the transformation decreases with
increasing Cu content and, at about 55% Cu, the transformation
becomes extremely sluggish. Inflection points were determined
which correspond to changes in the slope of the curves; these data
given in Table 1, show the transformation temperature (T_A) to be
~0.3, 0.4, and 0.5 Tm, where Tm is the average of the melting points
of Cu and Mn in °K. T_A was also found to decrease with increasing
Cu content.

Variation of the coefficient of thermal expansion (α) with
alloy composition is shown in Fig. 2. Throughout the compositional
range and between room temperature and 350°C, α is found to decrease
gradually with increasing Cu content. In the temperature range 350-
450°C, α decreases sharply up to 30% Cu and remains fairly constant
between 30 and 55% Cu. Over the range 450-550°C, α decreases up to
30% Cu and sharply increases beyond 30% Cu. In the range 550-650°C,
there is a sharp rise in α with increasing Cu up to about 55%. Above
650°C, α has a high value, which decreases as the Cu concentration
increases in the alloy.

All three transformations exhibit volume contraction, the magnitude varying with alloy composition and becoming insignificant at
about 55% Cu. The volume contraction, as calculated[4] from the
thermal-expansion data, is given in Table 2.

DISCUSSION

Mn exists in three allotropic forms, viz., γ, β, and α. Transformations from γ to β and finally to α exhibit negative volume
changes. Since Cu has negligible solubility in β- and α-Mn but
forms a series of solid solutions only in the γ-form of Mn, Mn-Cu
alloys retain the metastable γ-phase (solid solution of Cu in γ-Mn)

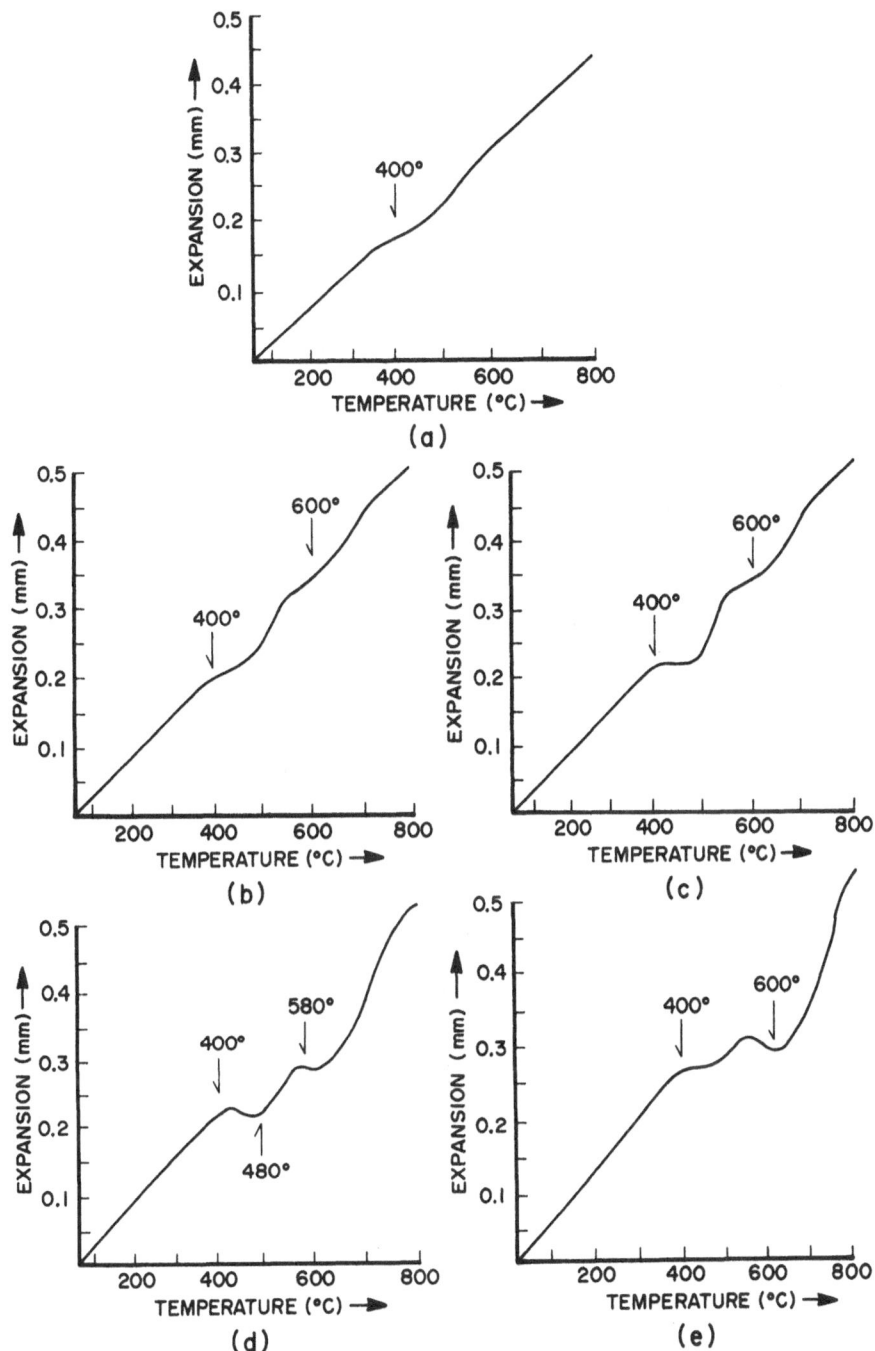

Figure 1.   Thermal-Expansion Curves for Mn-Cu Alloys: (a) Mn-55%
            Cu, (b) Mn-40% Cu, (c) Mn-35% Cu, (d) Mn-30% Cu, (e)
            Mn-15% Cu.

D. K. CHATTERJEE AND P. VENKATESWARARAO

Table 1.   Transformation Temperature (T_A)

Alloy Composition (wt% Cu)	Temperature (°C)		
15	340	400	600
20	380	480	590
35	350	450	560
40	340	425	530
45	320	400	520
55	300	450	525

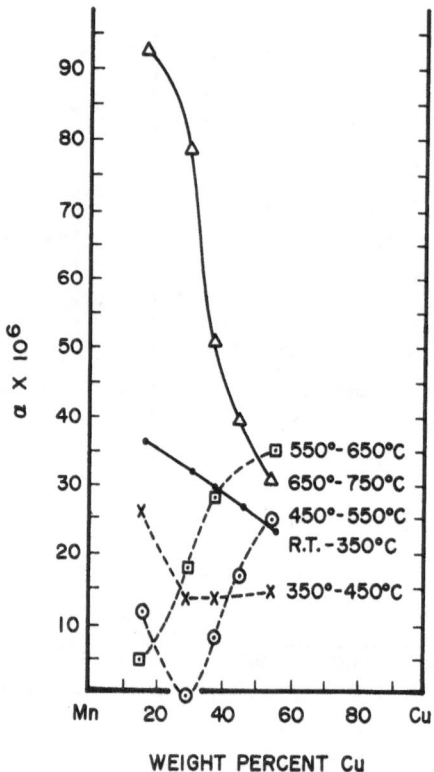

Figure 2.   Coefficient of Thermal Expansion vs. Composition Curves for Mn-Cu Alloys.

Table 2.   Volume Contraction (%) During Transformation

Alloy Composition (wt% Cu)	Volume Contraction (%)		
15	0.750	2.025	Negligible
30	1.410	1.800	Negligible
35	1.575	1.575	Negligible
40	1.425	1.200	Negligible
45	0.850	1.050	Negligible
55	0.750	Negligible	Negligible

upon quenching from 800°C. During heating of the quenched alloys, the first change in the slope of the theraml-expansion-vs-temperature curve is observed around 300°C (0.3 Tm), and volume contraction takes place during this heating. It is thought that these alloys, when heated to this temperature, begin to separate into two components--γ solid solution having a different composition than that of the initial matrix (i.e., γ-phase) and γ-Mn which is one of the components of the solid solution. This phenomenon is dependent upon the Cu content in the alloy because the critical points at which volume contraction commences are in the range 300-400°C (around 0.3 Tm) and depend upon the composition of the alloy (as shown in Tables 1 and 2). Moreover, from Fig. 2 it is apparent that thermal expansion in this range decreases sharply up to about 30% Cu and remains fairly constant with an increasing trend up to 55% Cu. Therefore, it is thought that the large volume contraction with negligible thermal expansion is responsible for the high magnitude of internal stresses in alloys having compositions around 40% Cu. At a specific composition of 40% Cu, the contraction effects which become significantly large mask the mild expansion of the γ-solid solution, while at compositions around 55% Cu and below 30% Cu, thermal-expansion effects are predominant. This explains the development of very high internal stresses in the matrix and why the highest damping capacity is observed[3] in Mn-Cu alloys at around 40% Cu concentration. Moreover, the coexistence of a γ-solid solution and γ-Mn as a possible single phase suggests that the transformation in the first stage may be a sort of intra-phase transformation, allowing an inter-growth of γ-phase and γ-Mn. This type of abnormal structure[5] may account·for the negative temperature coefficient of resistance[1,2] exhibited by the alloys in the 40%-Cu composition range.

However, in the other two steps of the transformations in the temperature ranges 400-500°C and 500-600°C, the unstable γ-Mn which separated in the first step transforms into β-Mn at about 500°C and

into α-Mn at about 600°C. These transformations seem to be substan-
tially complete with a sharp rise in thermal expansion in the range
650-750°C. The very high values of thermal-expansion coefficient
(α) in this temperature range are possibly due to the cumulative
effect of the thermal expansion of γ-solid solution and α-Mn.

From the above observations it can be concluded that quenched
Mn-Cu alloys retain γ-phase and are stable up to temperatures corresponding to 0.3 Tm. Transformations in these alloys commence at
about 300°C, and the alloys are continuously undergoing phase changes
in the temperature range 300-600°C and the transformations are sub-
stantially complete at about 650°C, resulting in two-phase alloys.
The various anomalous and interesting physical properties of the
γ-phase in Mn-Cu alloys are, therefore, transient effects at different stages of disintegration and formation of unstable intermediate structures prior to attaining equilibrium.

CONCLUSIONS

1. Quenched Mn-rich Mn-Cu alloys exhibit a single-phase structure up to 300°C and undergo a phase transformation in the temperature range 300-650°C before exhibiting two-phase structures: γ-solid
solution and α-Mn.

2. Quenched Mn-Cu alloys are metastable and undergo phase
transformations between 0.3 and 0.5 Tm in three steps, depending
upon the Cu concentration of the alloys. The first step (300-400°C)
is an intra-phase transformation where elemental γ-Mn separates from
the quenched γ-solid solution. The other two steps (at 400-500°C,
500-600°C) are associated with transformations of elemental Mn
($\gamma \rightarrow \beta \rightarrow \alpha$-Mn) in the matrix. The various abnormal and interesting
physical properties are transient phenomena which are dependent
upon the phase stability.

3. In the temperature range 300-400°C, large volume changes
and fairly steady values of the thermal-expansion coefficient (α)
in the compositional range 30-40% Cu give rise to large internal
stresses, causing these alloys to exhibit a high damping capacity.
The intra-phase transformation which allows inter-growth of γ-Mn
and γ-phase as a single phase may explain the negative temperature
coefficient of resistivity exhibited by these alloys.

REFERENCES

1. R. S. Dean, "Electrolytic Manganese and Its Alloys,"
        Ronald Press, NY (1952).
2. A. H. Sully, "Manganese," Butterworths, Woburn, MA (1955).

3. P. Venkateswararao and D. K. Chatterjee, Structural Studies
    of the Alloying Behavior of γ-Mn and Development of
    High Damping Capacity in Mn-Cu Alloys, J. Mat. Sci.
    15:139 (1980).
4. V. T. Cherepin and A. K. Mallik, "Experimental Techniques
    in Physical Metallurgy," Asia Publishing House, Bombay,
    (1967).
5. K. L. Chopra, "Thin Film Phenomenon," McGraw-Hill, NY
    (1969).

ITES SESSION 2: MEASUREMENT TECHNIQUES

Session Chairman: T. A. Hahn
                      National Bureau of Standards
                      Washington, DC

DEVELOPMENT OF A LASER INTERFEROMETRIC DILATOMETER*

W. D. Drotning

Sandia National Laboratories**
Albuquerque, New Mexico 87185

ABSTRACT

The ongoing development of a high precision dilatometer using laser interferometry is described. Design criteria require operation over a temperature range from ambient to 900 K with a length change sensitivity of ten microinches (0.25 μm) or less. In addition, the dilatometer is designed to achieve rapid sample turnaround for quality control measurements in a production environment. To achieve this, design features are incorporated which minimize sample preparation time by relaxing critical size and shape restrictions on test samples. The length change measurement is based on a modified Michelson interferometer using a two-frequency HeNe laser and ac detection of fringes. The final critical alignment of the interferometer during sample installation is achieved by automatic computer control. This alignment control feature can be used to maximize interferometer alignment continuously through the course of a run. The furnace system incorporates a low thermal mass heating chamber to achieve rapid thermal response with minimal thermal disturbance to the interferometer system. Flexibility and reproducibility in heating schedules are achieved through digital control. Dilatometer control and data acquisition functions are accomplished by a minicomputer using a standard interface bus. Areas for further development are also discussed.

---

*This work was supported by the U.S. Department of Energy (DOE), under Contract DE-AC04-76-DP00789.
**A U.S. DOE facility.

INTRODUCTION

There is a need for routine high temperature dilatometric
measurements with a precision greater than can be obtained by conventional push-rod dilatometry. An example of this need is the
matching of thermal expansion coefficients of glass-ceramic and
metal components used in sealing applications. This paper describes the development of a prototype high precision laser interferometric dilatometer, which is suitable for operation in a
production environment where quality control data is required on
a routine basis.

To the best of our knowledge, this is the first interferometric dilatometer which has been designed to operate in a non-
laboratory environment with high sample throughput. A literature
survey of interferometric dilatometry is given in Ref. 1, which
indicates that previous devices in this field have been dedicated
solely to the research laboratory. The operation of such dilatometers is characterized by long sample measurement time and tedious
sample preparation and fixturing, requiring highly-skilled person-
nel for both assembly and operation. The goal of this development
effort is a dilatometer with the precision afforded by laser interferometry, the operational speed and convenience of push-rod dilatometry, and state-of-the-art automatic control, data acquisition,
and analysis.

Dilatometer design and performance criteria are presented in
the next section, followed by a detailed discussion of the technical design. The length change measurement is based on a
modified Michelson interferometer, while the furnace and sample
are arranged in a vertical configuration. (Details of the design
selection are given in Ref. 1.) A unique automatic optical alignment feature is also presented. Finally, areas for further
development are discussed.

DILATOMETER DESIGN AND PERFORMANCE CRITERIA

The following criteria were established for the design and
performance of the high precision dilatometer: operation over the
temperature range from ambient to 900 K in a vacuum or inert gas
atmosphere with a sample length change measurement sensitivity of
10^{-5} in. (2.5×10^{-4} mm) or less. For a one inch sample and a
temperature change of 600 K, this sensitivity corresponds to a
measurement precision of the mean thermal expansion coefficient of
$\Delta\ell/(\ell\Delta T) \sim 2 \times 10^{-8}$/K. The dilatometer should be operational for
a variety of materials covering a wide range of expansion coef-
ficients, at least from 10^{-4}/K to 5×10^{-7}/K.

While such a dilatometer would be useful as a research laboratory instrument, a number of additional requirements were established in order to achieve the optimistic design goal of measurement of two samples per day in a production environment. First, both sample size and shape should be non-critical, in order to minimize sample preparation time. Second, sample measurement time should be minimized, by incorporating design features which facilitate ease of operation with a minimum of set-up time. Critical optical alignment procedures should be minimized, or automated if possible. Automatic data acquisition and reduction should also be used. In short, for a quality control application, the dilatometer design should allow high precision thermal expansion measurements to be made on a routine basis. To aid in the current development, commercially-available equipment should be incorporated wherever possible.

TECHNICAL DESIGN

A schematic diagram of the laser interferometric dilatometer system is shown in Fig. 1. The IEEE-488 interface bus has been adopted for all communication between the dilatometer and a minicomputer-based data acquisition system. The dilatometer system consists of the following components:

Optical Interferometer

A detailed schematic of the optical interferometer is shown in Fig. 2. The laser/detector consists of a Hewlett-Packard 5526A Laser Measurement System which uses a two-frequency HeNe laser to achieve heterodyne frequency mixing of the phase information in each arm of the double-beam interferometer. The interferometer itself is based on the Michelson design, and has been specifically adapted for use with the two-frequency laser system. The heterodyne mixing technique allows ac detection of fringes in the frequency domain, and has several advantages over conventional dc (single frequency) fringe counting techniques.[1] The laser output consists of two frequencies, f_1 and f_2 ($\approx 4.7 \times 10^8$ MHz), separated by approximately 2 MHz, and coded by their orthogonal linear polarizations. The two frequencies are split into the two optical paths shown in Fig. 2 by a polarizing beamsplitter (PB) and quarterwave retarders, so that each optical path into the furnace represents a single frequency of laser light. The two beams reflect from front-surface mirrors in the furnace which are separated by a distance equal to the sample dimension. Upon expansion or contraction of the sample, the interference pattern detected by the laser system is varied by an amount proportional to the change in sample length. The laser interferometer system directly measures and displays this change in length, both in sign and magnitude.

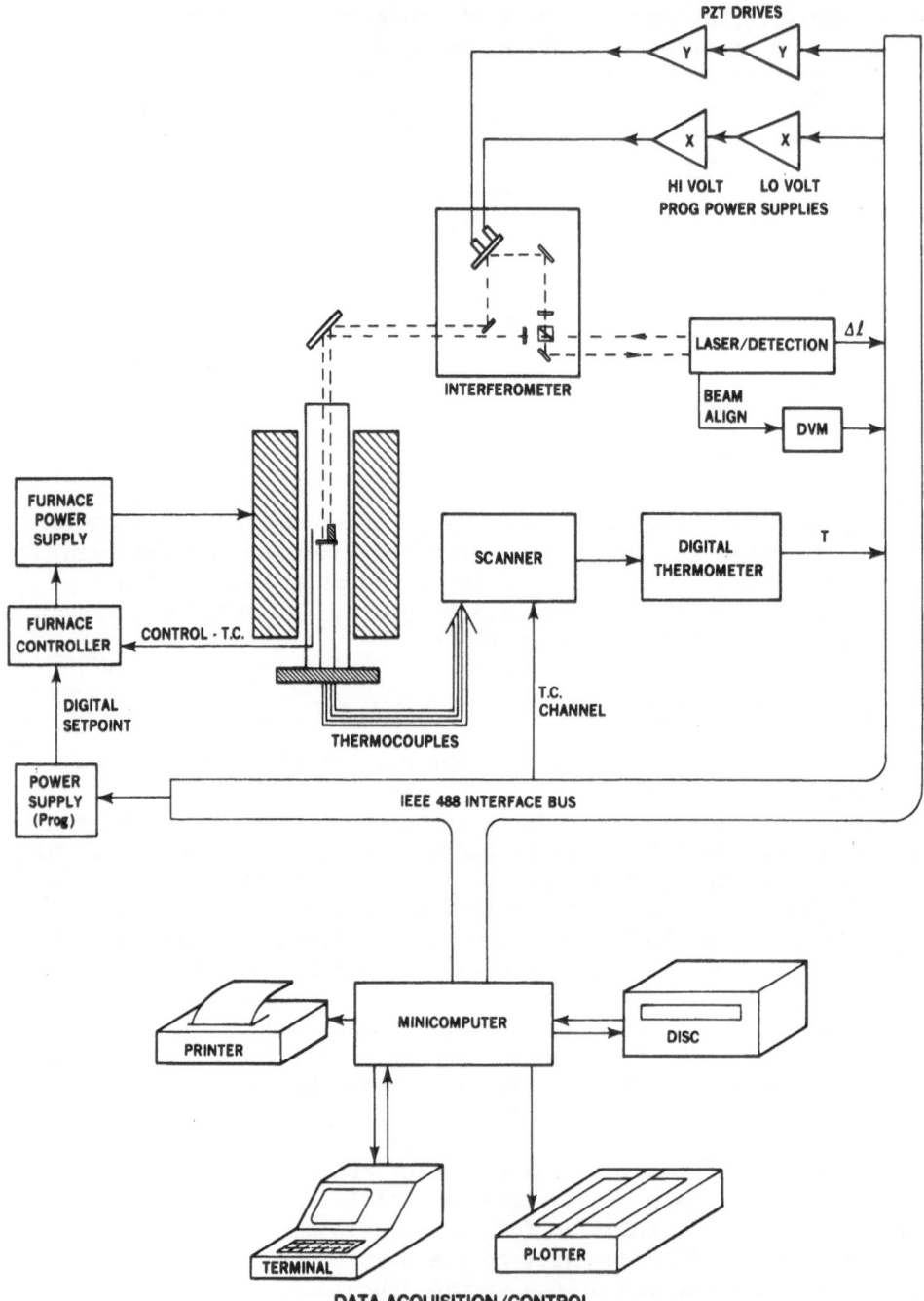

Figure 1. Schematic diagram of the laser interferometric dilatometer. Optical paths are shown as dashed lines and
electrical signal and control paths by solid lines.

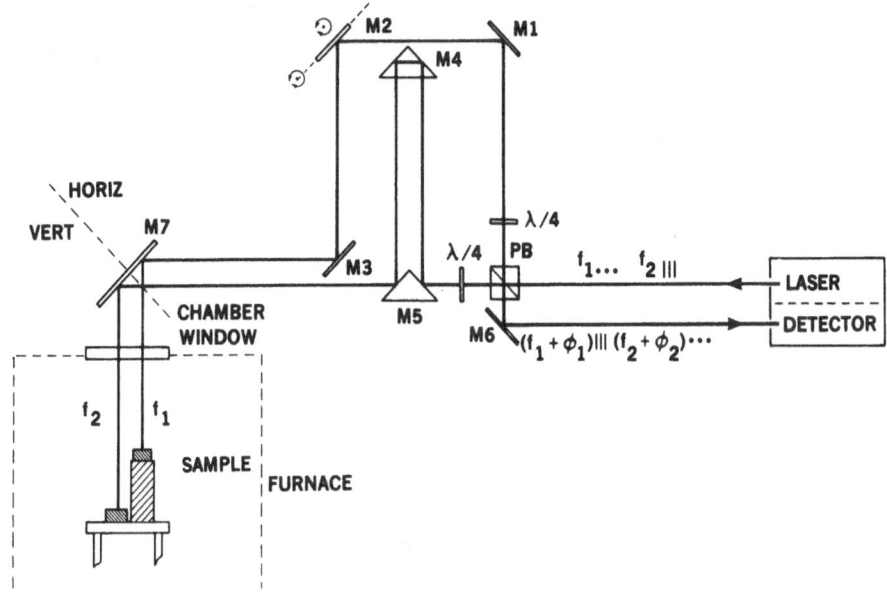

Figure 2. Detailed schematic diagram of the optical interferometer portion of the dilatometer. The interferometer
is based on the double-beam Michelson design, using a
two-frequency laser. The frequencies f_1 and f_2 are
coded by their orthogonal linear polarizations ($\cdot$ and $|$).
The return signals, also linearly polarized perpendicular to each other, are heterodyne frequency mixed in
the detector.

The mirrors M4 and M5 form an optical path which corrects for
thermal expansion of the steel optical table on which the optics
are mounted. Mirror M7 directs the beams from the horizontal
plane of the laser, detector, and optics on the table to the vertical plane for entry into the furnace. Mirror M2, in the sample
optical beam, is mounted in a two-axis gimbal mount, which allows
for adjustment of the beam direction onto the sample and, on
return, to the detector. This adjustment is necessary to compen-
sate for variations in sample surface flatness and parallelism.
Both manual and electronic adjustment of this mirror are possible
through micrometer screw drives and electronically controlled
piezoelectric transducers (PZT).

Auto-Alignment

The interferometer is automatically aligned through the feed-
back circuit depicted in Fig. 1. A beam alignment signal is

extracted from the laser detector system and sent to the minicomputer. If the alignment falls below a preset value, indicating
degradation of the alignment of the interferometer, the PZT drives
are automatically varied to return the beam alignment signal to a
maximum value. This is accomplished by programmatically stepping
a low voltage DC power supply, which drives a programmable high
voltage DC supply, which in turn drives the PZT. During heating,
thermal gradients and differential expansion can tend to misalign
the optical beams; if the alignment becomes too degraded, the
fringe pattern is destroyed, and length change information will be
lost. The auto-alignment feature can be used throughout the run
to prevent this, and thereby allows more rapid heating rates than
could otherwise be used. Auto-alignment may also be used at the
beginning of a run to assist the operator in the final critical
alignment of the interferometer.

Sample Mirrors

The sample and reference mirrors used in the furnace are 12
mm diameter x 6 mm thick fused silica mirror blanks, with a
multi-layer dielectric coating for maximum reflectance (>99%) at
0° incidence, 632.8 nm wavelength. These mirrors have been successfully cycled from ambient to 900 K several times without
serious degradation.

Furnace System

The furnace is a radiant heating chamber (Research Inc.,
Minneapolis, MN) which achieves its low thermal mass by using
water-cooled polished aluminum elliptical mirrors to reflect
radiation from four quartz lamps toward the furnace center. A
horizontal cross-section of the furnace and sample enclosure is
shown in Fig. 3. The split-shell furnace design offers easy
access to the sample area in a vertical configuration. The low
thermal mass allows rapid furnace response and cooldown, thus
improving sample turnaround time. In addition, the water-cooled
furnace jacket minimizes thermal disturbances to the atmosphere
surrounding the furnace, thus increasing interferometer stability
by reducing convective air currents and heat load to the optics
and mountings. The sample is enclosed in a quartz retort tube,
which is evacuated by a mechanical pump. The cylindrical sample
enclosure is made from graphite, and will be described later.

In order to reduce sample set-up time, the sample temperature is not measured directly. Instead, the sample enclosure
temperature, which is measured using type K or S thermocouple
thermometry, is used to infer the sample temperature by application of known corrections. This also eliminates wires attached
directly to the sample which tend to cause sample movement and
destroy optical alignment of the interferometer.

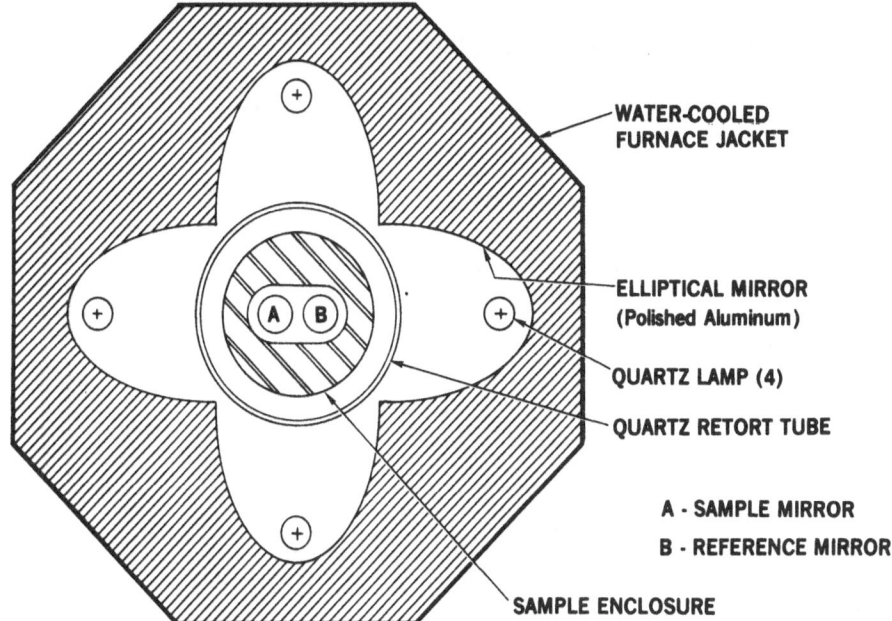

WATER-COOLED
FURNACE JACKET

ELLIPTICAL MIRROR
(Polished Aluminum)

QUARTZ LAMP (4)

QUARTZ RETORT TUBE

A - SAMPLE MIRROR

B - REFERENCE MIRROR

SAMPLE ENCLOSURE

Figure 3.   Horizontal cross-section of dilatometer furnace,
sample enclosure, and evacuated retort tube.

Sample Support

     An exploded vertical cross-section of the sample support is
shown in Fig. 4. The sample is placed on a base of polished
silicon, which is held on the fused silica base and support by a
graphite clamp. Graphite is used because of its high absorptance
of the quartz lamp radiation, while fused silica is used for its
low absorptance, low thermal conductivity, and low thermal expansion. The silicon base offers a high thermal conductivity for
good thermal equilibration, yet low thermal expansion. The
sample itself is nominally 2 cm long x 8 mm diameter, with ends
flat, parallel, and polished. The sample and reference mirrors
are sheathed with thin graphite collars to improve thermal equilibration. The sample and mirrors are contained in the graphite
sample enclosure which rests on the clamp. The lid has two apertures for entry of the beams. Optical access to the evacuated
retort tube is through an anti-reflection (AR) coated window,
which is water-cooled by its mount. The radiation shields above
the AR window consist of polished aluminum surfaces to reflect
radiant heat to the sample enclosure and optical beam apertures
(8 mm diameter). These apertures and reflecting surfaces greatly
reduce the thermal load on the optical mount situated just above

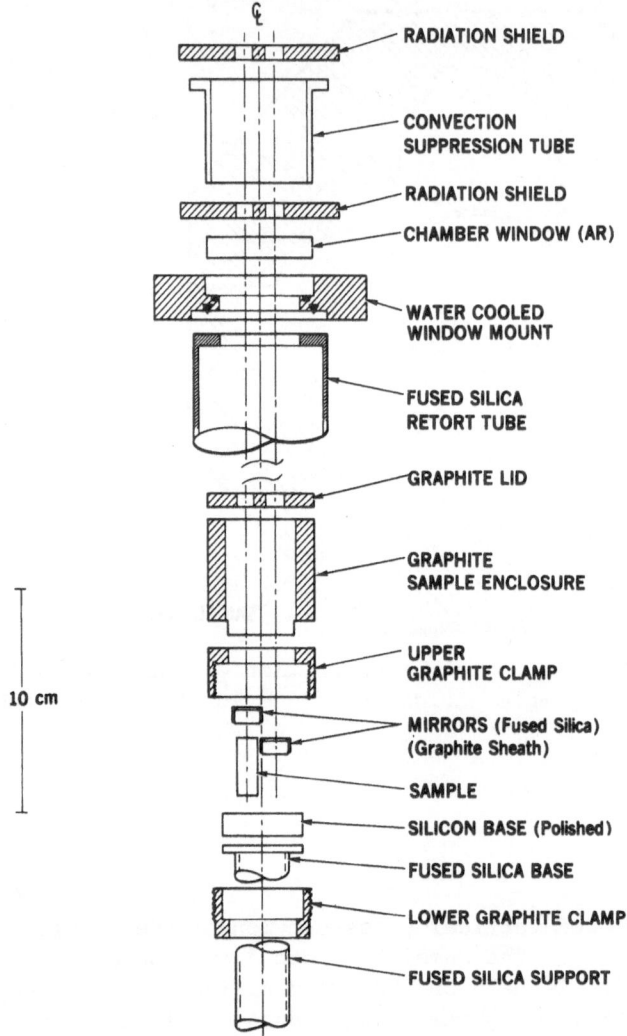

RADIATION SHIELD

CONVECTION
SUPPRESSION TUBE

RADIATION SHIELD

CHAMBER WINDOW (AR)

WATER COOLED
WINDOW MOUNT

FUSED SILICA
RETORT TUBE

GRAPHITE LID

GRAPHITE
SAMPLE ENCLOSURE

UPPER
GRAPHITE CLAMP

MIRRORS (Fused Silica)
(Graphite Sheath)

SAMPLE

SILICON BASE (Polished)

FUSED SILICA BASE

LOWER GRAPHITE CLAMP

FUSED SILICA SUPPORT

10 cm

Figure 4. Vertical cross-section (exploded view) of the sample
support and heating fixtures. Components external to
the furnace and evacuated retort tube are also shown.

the top shield. The shields are separated by a tube which sup-
presses vertical convective currents which would degrade the
interferometer fringes.

The entire sample support and enclosure mechanism has been
empirically designed to achieve thermal uniformity of the sample
and base as observed by direct measurements of sample temperature
gradients and indirectly, yet more sensitively, by measurement of
the optical interferometer alignment response at different temperatures.

Data Acquisition and Control

All data acquisition and control functions are accomplished by a Hewlett-Packard HP 1000 minicomputer system, which communicates with the dilatometer via the IEEE-488 interface bus (see Fig. 1). The minicomputer performs the following functions:

Furnace Control. All furnace heating schedules are controlled by a user-defined series of heating/cooling ramps and isothermal holds.

Data Acquisition. Temperature data from a digital thermometer and expansion data from the laser measurement system are read periodically and are stored on a random access disc for subsequent analysis.

Interferometer Alignment. The beam alignment is continuously monitored and sent to the computer. If the alignment signal falls too low, a stepping sequence is automatically begun on the two PZT drives in order to return the alignment signal to the maximum value. Special error conditions result if the signal cannot be maximized, if the PZT drives are out of range, or if the alignment signal is too low.

Remote Real-Time Status. The status of the dilatometer can be continuously monitored at a remote site. This includes information regarding the sample under test, the heating schedule being used, the current temperature and expansion values, data from the auto-alignment system, and the status of the minicomputer control programs.

Analysis and Display. Following a run, the stored data are analyzed and can be sent in raw or reduced form to a printer and a graphics plotter. An example of the latter is shown in Fig. 5, where expansion x is plotted vs temperature for a sample of glass-ceramic, designated MS011A. Deviation from non-linearity is due to thermal lag between sample and measured temperatures. Since the abscissa is temperature and not time, the isothermal hold which was achieved at the highest temperature is not readily apparent. The spike in the data near 420°C is an intentional marker, and indicates the occurrence of an auto-alignment. The data summary at the right of the plot gives a description of the run for later reference.

FUTURE DEVELOPMENT AREAS

Several areas which require further development effort have been identified. These include:

Minimization of Thermal Gradients

Currently, we have obtained vertical thermal gradients in vacuum of less than 0.2 K/cm on low conductivity samples (fused silica) at 900 K. Work is continuing to affect design modifications to reduce these gradients. Investigation of any temperature

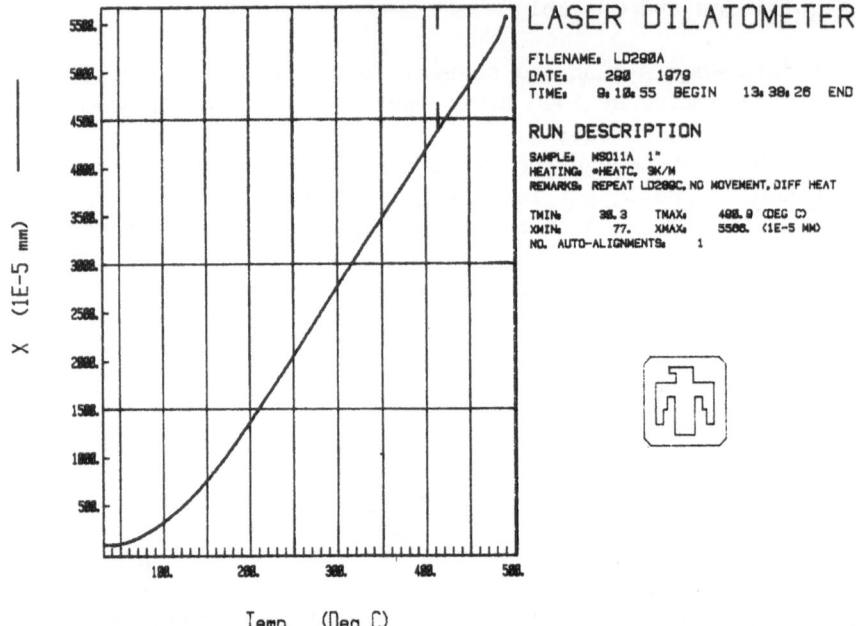

Figure 5.   Typical output of raw data from the graphics plotter for
            a sample of glass-ceramic, designation MS011A.

difference between the sample and reference mirrors is also continuing.

Sample Temperature Measurement

We are investigating methods to increase the precision of sample
temperature measurements while not directly contacting the sample.
This includes installation of calibrated type S thermocouples, and
careful measurements of sample and enclosure temperatures with
samples of various conductivities and emittances in order to infer
the sample temperature by correlation.

Standards Testing

The next major phase of development will include testing and
calibration using NBS standard reference materials (copper, fused
silica, sapphire, and borosilicate glass). In addition, we are
planning round-robin tests with push-rod dilatometers to establish
internal standards at our laboratories.

Define Sample Specifications

In order to use the dilatometer on a routine basis, specifications for sample preparation need to be defined. Besides gross sample sizes, this includes specifications for the allowable tolerances on surface flatness, parallelism, and polish.

Establish Operational Procedures

Detailed procedures need to be specified for dilatometer operation. Several modes of operation are available, including total sample elongation while ramping temperature, total elongation at a series of isothermal holds, and sample CTE (coefficient of thermal expansion) at a series of isotherms.

CONCLUSION

The design and performance criteria for a laser interferometric dilatometer suitable for routine operation were established, followed by a detailed description of the major components of the dilatometer design. The double-beam optical interferometer utilizes a two-frequency laser for ac fringe detection. A computer-controlled auto-alignment feature is employed to enhance operation. A low thermal mass furnace is used in a vertical geometry for sample heating. Details of the automatic control, data acquisition, and analysis functions are also described. Current problems and future development areas are discussed in the final section. The development phases yet to be accomplished include standards testing and calibration, definition of sample specifications, and establishing operational procedures.

REFERENCE

1. W. D. Drotning, "Design of a High Precision Dilatometer Using Laser Interferometry," SAND 78-1796, Sandia Laboratories, Albuquerque, NM, (December, 1978). Available from: National Technical Information Service, U.S. Dept. of Commerce, 5285 Port Royal Rd., Springfield, VA 22161.

A LOW TEMPERATURE (4.2 to 350 K) DIFFERENTIAL DILATOMETER

C. A. V. de A. Rodrigues, M. Carrard, J. Plusquellec,
and P. Azou

Institut de Physique et de Métallurgie
Ecole Centrale des Arts et Manufactures (92290)
Châtenay-Malabry, France

ABSTRACT

A new tube-type vitreous silica vertical differential dilatometer for the measurement of the coefficient of thermal expansion
of solids is described. Thermal dilatation is measured by means of
an inductive displacement tranducer (LVDT) and its electronic equipment. The present detection limit of the apparatus, due to the
analog recorder, is 1×10^{-4} mm ($\Delta\ell/\ell = 2 \times 10^{-6}$) for a maximum
differential expansion $\Delta\ell$ ($\Delta\ell = \Delta\ell_{sample} - \Delta\ell_{reference}$) of $\pm 3 \times 10^{-1}$
mm, or 3.33×10^{-4} mm for a maximum $\Delta\ell$ of ± 1 mm (useful range). The
maximum recorded drift rate was 2.5×10^{-4} mm/24 hours. The force
applied by a vitreous silica push rod on the sample can be modified
from 0 to 1N even during an experiment. A liquid helium circulation
cryostat and a furnace with its associated electronic equipment are
employed concurrently during experiments in order to achieve linear
cooling and heating at rates as low as 0.04 K/min and as high as
5 K/min, as well as isothermal maintenance of the sample and the
reference in the 4.2 to 350 K temperature range. The above specifications make this apparatus especially useful to study low temperature phase transformations. In fact, the apparatus is presently
being used to study the austenite to martensite transformation of
Fe-Ni-C alloys which have a low Ms temperature (below 240 K), as
well as connected phenomena. Results on a Fe-24.2%Ni-0.41%C
(weight percent) alloy are presented.

INTRODUCTION

Measurements on the thermal expansion of solids at low temperatures have yielded considerable information concerning various

67

phenomena that occur in solids. Consequently, there has been an
effort to develop new types of dilatometers, as well as to perfect
existing types in order to increase their sensitivity and to facilitate their operation.

For a better understanding of the austenite to martensite
transformation of Fe-Ni-C alloys which have a low Ms temperature
(below 240 K), as well as connected phenomena, it is desirable to
extend the measurements of their expansion $(\Delta \ell)$ to liquid nitrogen
and to liquid helium temperatures. The present paper describes the
details of a new tube-type vitreous silica differential dilatometer
recently developed in our laboratory for the above applications.

APPARATUS

A general view of the vertical-type low temperature differential dilatometer consisting of the dilatometer head and system,
the cryostat, and the support frame is shown in Figure 1.

Dilatometer Head

The dilatometer head is shown in Figure 2. The present detection limit of the apparatus, due to the analog recorder resolution,
is 1×10^{-4} mm for a maximum differential expansion $\Delta \ell$ ($\Delta \ell = \Delta \ell_{sample}$
$-\Delta \ell_{reference}$) of $\pm 3 \times 10^{-1}$ mm, or 3.33×10^{-4} mm for a maximum $\Delta \ell$ of
± 1 mm (useful range). The apparatus has a sensitivity to relative
length changes $\Delta \ell / \ell$ of 2×10^{-6}. The maximum recorded drift rate
of the apparatus was 2.5×10^{-4} mm/24 hours. These values are considered acceptable for our purposes.

Two inductive displacement transducers (noted 1 and 2 in Figure
2) or LVDTs (Linear Variable Differential Transformer) and their
associated electronic equipment (Amplifier 1 and 2, respectively)
are used: LVDT 1 for the measurement of the differential expansion;
and LVDT 2 for the measurement of the force exerted on the sample
by its push rod. The gain of Amplifier 1 can be selected from
$(1 \rightarrow 999) \times 10$. Furthermore, the applied force can be modified from
0 to 1N by means of an electric motor (3) even during an experiment.
The transducers (1 and 2) are mounted on movable stages (4 and 5,
respectively) that roll on linear ball bearings (not shown). Because of the vertical disposition of the measuring head, 4 and 5
have a weight compensating device (6 and 7, respectively). Thermal
stabilization of the measuring head is achieved by a thermostatically controlled circulator which maintains distilled water at
293 ± 0.1 K (see water circuit 8).

An external gas circuit allows control of the atmosphere in a
chamber through orifice 9 situated on top of the aluminum enclosure
of the head (10). It is possible to realize a primary vacuum, a

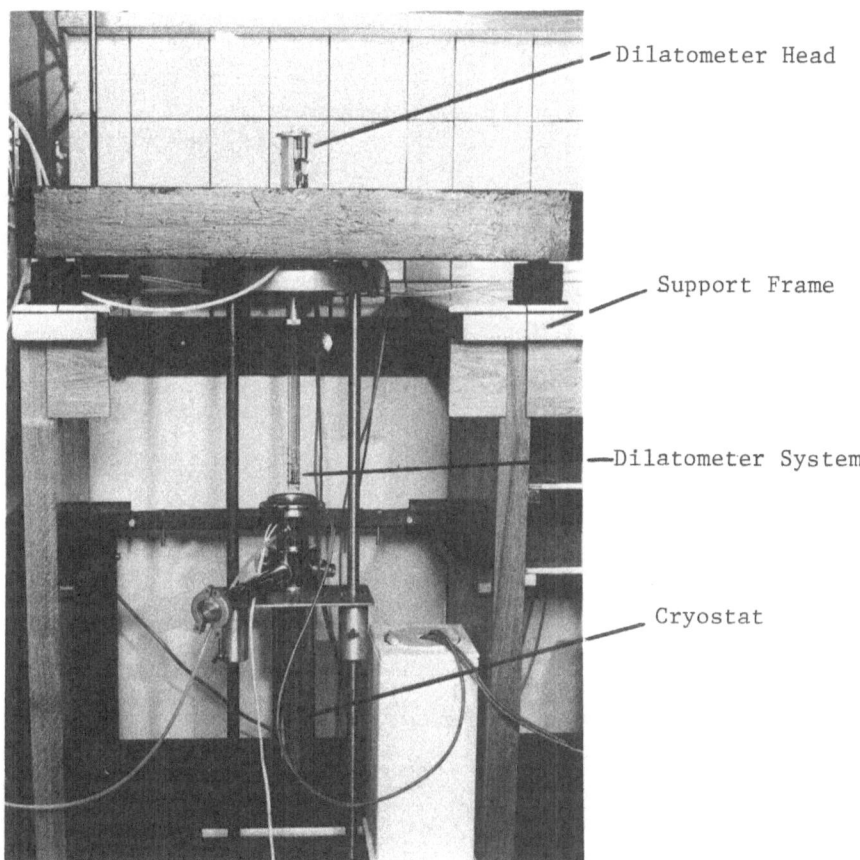

Figure 1. General view of the low temperature dilatometer.

1 – LVDT 1 ($\Delta\ell$)
2 – LVDT 2 (F)
3 – Electric Motor
4 – Movable Stage ($\Delta\ell$)
5 – Movable Stage (F)
6 – Weight Compensating
    Device ($\Delta\ell$)

7 – Weight Compensating
    Device (F)
8 – Water Circuit
9 – Orifice (vacuum)
10 – Aluminum Enclosure of the Head
11 – Orifice (helium gas)

Figure 2a.   Dilatometer head.

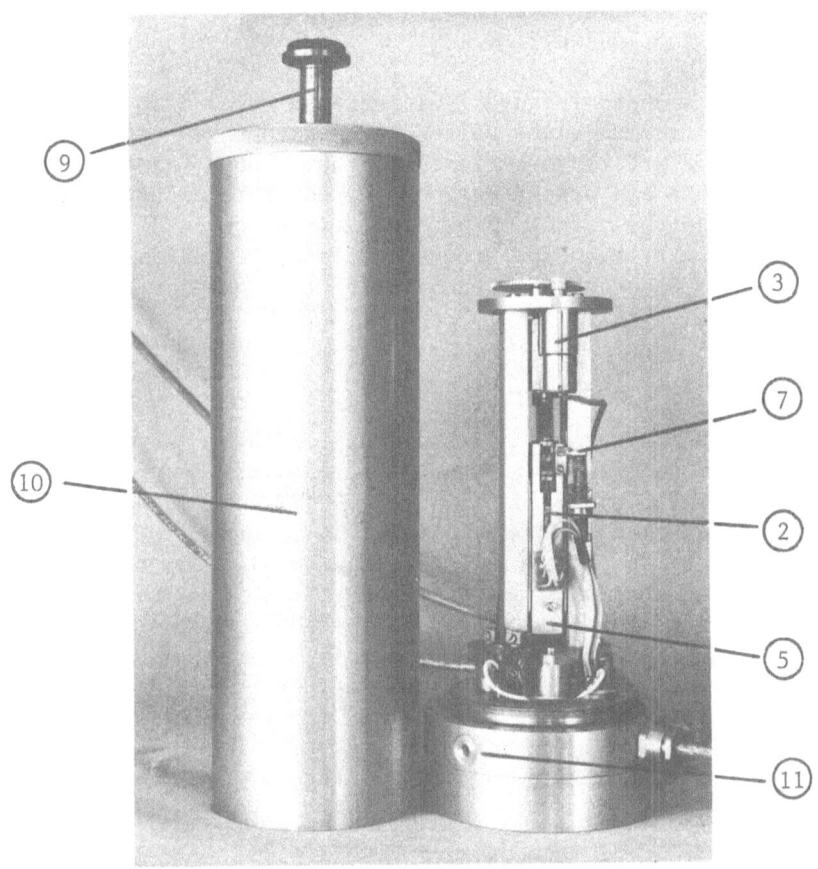

1 - LVDT 1 ($\Delta\ell$)
2 - LVDT 2 (F)
3 - Electric Motor
4 - Movable Stage ($\Delta\ell$)
5 - Movable Stage (F)
6 - Weight Compensating
    Device ($\Delta\ell$)

7 - Weight Compensating
    Device (F)
8 - Water Circuit
9 - Orifice (vacuum)
10 - Aluminum Enclosure of the Head
11 - Orifice (helium gas)

Figure 2b. Dilatometer head.

static atmosphere, or a dynamic atmosphere in the chamber. Generally, helium gas, piped through another orifice (11) into the chamber, is used as an exchange gas for cooling and heating of the sample and the reference. The above chamber consists of the following parts: the upper part of the chamber is the aluminum enclosure of the head (10), and the lower part is a metal protection tube that extends into the cryostat (not shown).

The dilatometer head (Figure 2) is screwed to a 30 mm thick, 300 mm diameter duralumin plate bolted below a heavy concrete triangular-shaped platform which rests on the support frame. Three pneumatic insulators are placed between the concrete platform and the frame to reduce the effect of external vibration on $\Delta\ell$ measurements (Figure 3). The resonant frequency of the mechanical mounting is about 3 Hz and leads to an attenuation of the dominant external frequencies of about 97%.

Dilatometer System

The dilatometer system consists of the following vitreous silica parts shown in Figure 4: a support tube (12), support platform (13), and two push rods (14). A 300 mm long support tube is fixed by a thin layer of epoxy to an aluminum sleeve (15) which is firmly screwed to the measuring head (Figure 4). The removable platform on which the sample and reference rest has two through-holes for thermocouples and is placed at the lower extremity of the support tube. Each push rod, fixed by a thin layer of epoxy to an eccentrically bored cylindrical aluminum rod, is screwed to a movable stage. The eccentricity of the bore allows for the alignment of the sample (or the reference) with its push rod. The length of the sample/reference push rod is a function of the sample/reference length.

A sample and a reference having flat and parallel ends, either the same size or not, up to 52 mm in length and 7 mm in diameter, can be mounted into the dilatometer system through a window in the support tube (Figure 4).

Cryostat-Furnace

Cooling to 4.2 K is achieved by pumping helium from a liquid helium container in and out of the circulation cryostat. The sample and the reference are isolated from the cryogenic fluid within the cryostat by the metal protection tube.

To heat the sample and the reference, a unifilar heating element was wound around and to the lower part of the protection tube in such a way that the generation of a magnetic field is avoided. To minimize temperature gradients along the sample and the reference,

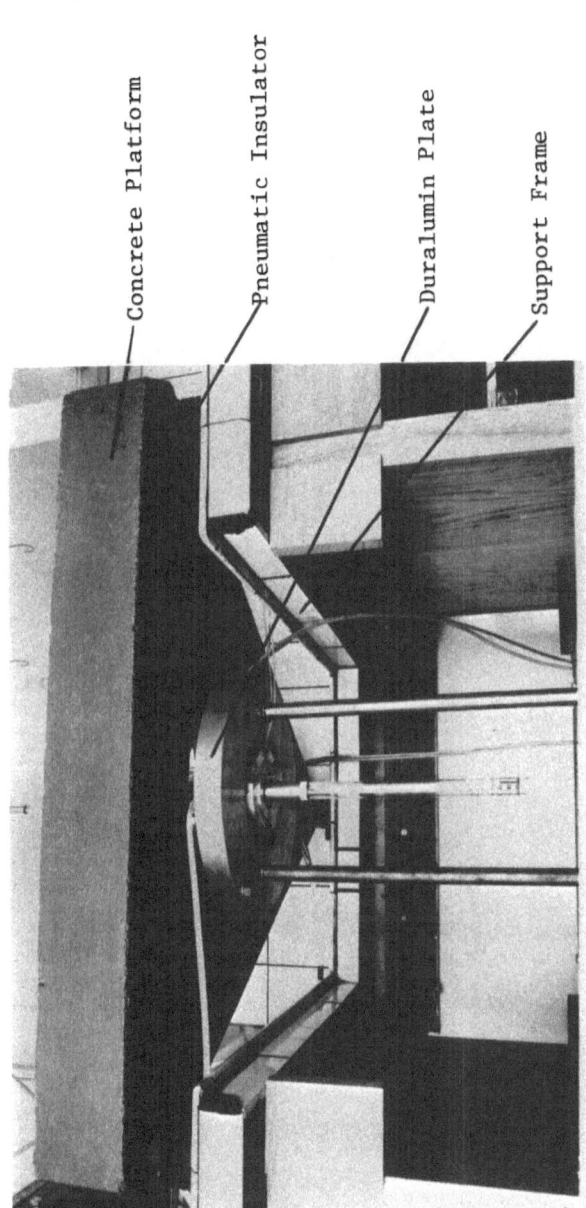

Figure 3.   View of the mechanical mounting of the dilatometer head.

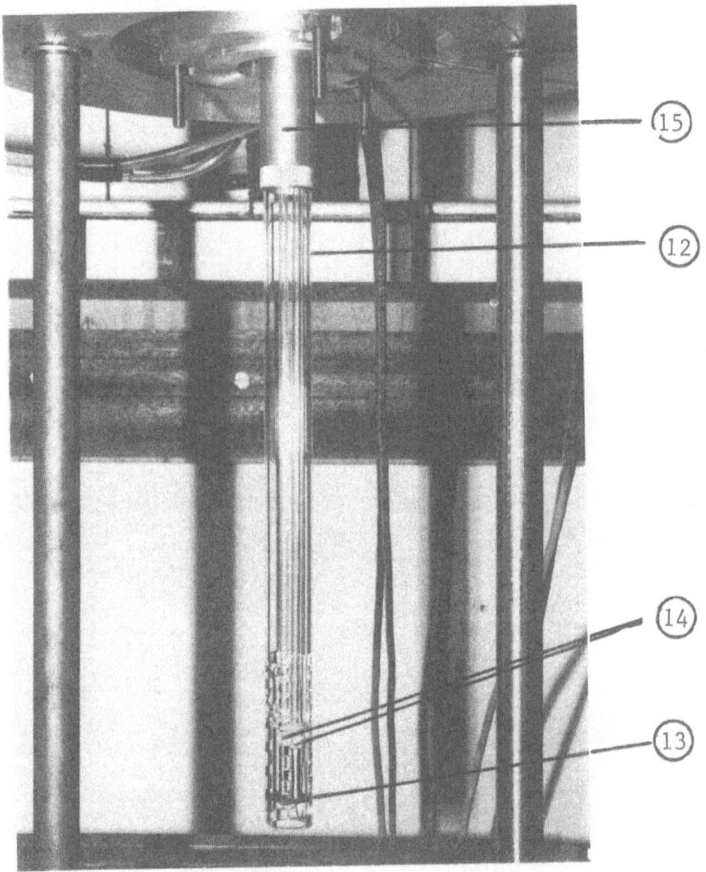

12 - Support Tube
13 - Support Platform
14 - Push Rods
15 - Aluminum Sleeve

Figure 4. Dilatometer system.

the lower part of the protection tube (furnace) is made of copper.
The upper part is made of stainless steel to avoid excessive thermal
conduction between the furnace and the measuring head. The cryostat
and the furnace with its associated electronic equipment (Figure 5)
are employed concurrently during experiments in order to achieve
linear cooling and heating, as well as isothermal maintenance of the
sample and the reference in the 4.2 to 350 K temperature range. The
furnace's electronic equipment consists of: a temperature "pro-
gramming" unit which includes the programmable dc voltage generator
and the furnace temperature sensor; a PID (Proportional, Integral,
Derivative) temperature regulator unit; and a tyristors power supply
unit. Due to the construction technology of the cryostat, temperatures above 350 K and below 4.2 K cannot be reached while using the
cryostat. Higher temperatures (about 500 K) should be attained if
the cryostat is removed. As shown in Figure 5 (schematic of apparatus), the temperature of the furnace is sensed by a gold-iron
versus chromel thermocouple and the set-point temperature of the
furnace (in volts) is set by the programmable dc voltage source
which delivers a dc-volt signal $V = f(t)$ which is a linear function
of time.

Cooling or heating the sample (s) and the reference (r) at a
fixed rate and measuring their temperature (Ts, Tr), as well as
thermal differential expansion ($\Delta\ell$) dynamically is not as accurate
as measuring Ts, Tr, and $\Delta\ell$ statically, i.e., by heating and cooling in a stepwise manner and allowing sufficient time for thermal
equilibrium of s and r at different temperatures to be established.
Despite this, both procedures are necessary when studying phase
transformation phenomena and can be achieved by our apparatus.
Generally, linear cooling and heating of the sample and the reference
between 4.2 and 350 K at rates from 1 to 2 K/min are performed.
These cooling and heating rates seem to be a good compromise between
1-day term experiments and between a small temperature deviation
between the sample and the reference, as well as small temperature
gradients along the sample and the reference. Nevertheless, cooling
and heating rates as low as 0.04 K/min and as high as 5 K/min are
achievable by our apparatus.

Thermometry

As shown in Figure 5, a digital thermometer with an analog
output was used for measuring the temperature of the specimen which
was sensed by a chromel-constantan (Type E) thermocouple inserted
into the center hole (4 to 6 mm deep) made in the lower extremity
of the specimen. This thermometer could only measure temperatures
down to 21 K, but since linear cooling can be achieved down to 4.2 K,
the lowest temperature attained was estimated to be 6 K. Two gold-
iron versus chromel thermocouples, calibrated throughout the temperature range of 4.2 to 300 K, are presently being used to sense the
temperature of the sample and of the reference.

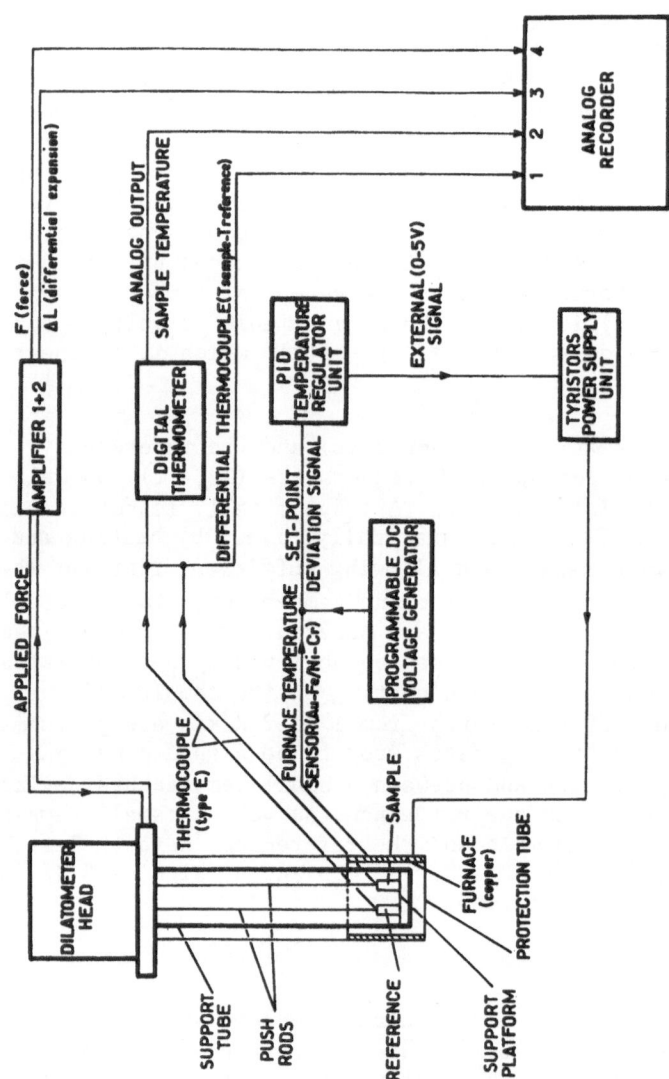

Figure 5.   Schematic of the apparatus.

RESULTS

Experimental Procedure

The apparatus described above was used to study a Fe-Ni-C alloy. Cylindrical specimens, 20 mm in length and 4 mm in diameter, were set up at room temperature before in-situ quenching to 77 or 4.2 K. The cooling rate was 1 K/min.

Material

The Fe-Ni-C alloy used had the following chemical composition: 0.41% C, 24.2% Ni, 0.27% Mn, 0.33% Si, 0.11% P, and 0.028% S (weight percent). After high temperature forging, an austenitizing treatment (1323 K/2 hours), followed by water quenching was performed on bars which had a cross section larger than the specimen size. After machining, the samples were mechanically and electrolytically polished and an entirely austenitic structure was obtained at room temperature. During cooling from Ms (martensite start) to Mf (martensite finish), the face-centered cubic (f.c.c.) austenitic structure transforms to a body-centered tetragonal (b.c.t.) martensitic structure. For example, after cooling to 77 K a mixed structure containing about 85% of martensite and about 15% of retained austenite (percent determined by metallography) is obtained. Results concerning only two samples (A and A') of the above Fe-Ni-C alloy are reported here.

Experimental Results

The differential dilatometric behavior of sample A and A' is plotted versus temperature in Figure 6 (Curves a and b) and Figure 7 (a' and b'), respectively. Curve a is for the initial cool down of Sample A from room temperature to 77 K. After a 90-hour aging at 77 K (curve not reported in this paper), Sample A was heated from this temperature to room temperature (Curve b). Curve a' is for the initial cool down of Sample A' from room temperature to 6 K. After a 15 minute aging at 6 K, Sample A' was heated from this temperature to room temperature (Curve b').

Curves a and a' show the contraction of the austenitic phase for temperatures above Ms (T>Ms) and the expansion associated with the austenite to martensite athermal transformation for T ≤ Ms (Ms temperature is close to 212 K). Curve a' also shows that athermal martensite still forms at temperatures below 77 K.

The exothermic character of the austenite to martensite transformation of Fe-Ni-C alloys can be observed in Figure 8 in which a typical evolution of the temperature deviation between a Fe-Ni-C sample and a reference $\Delta T = Ts-Tr$ is plotted versus the temperature

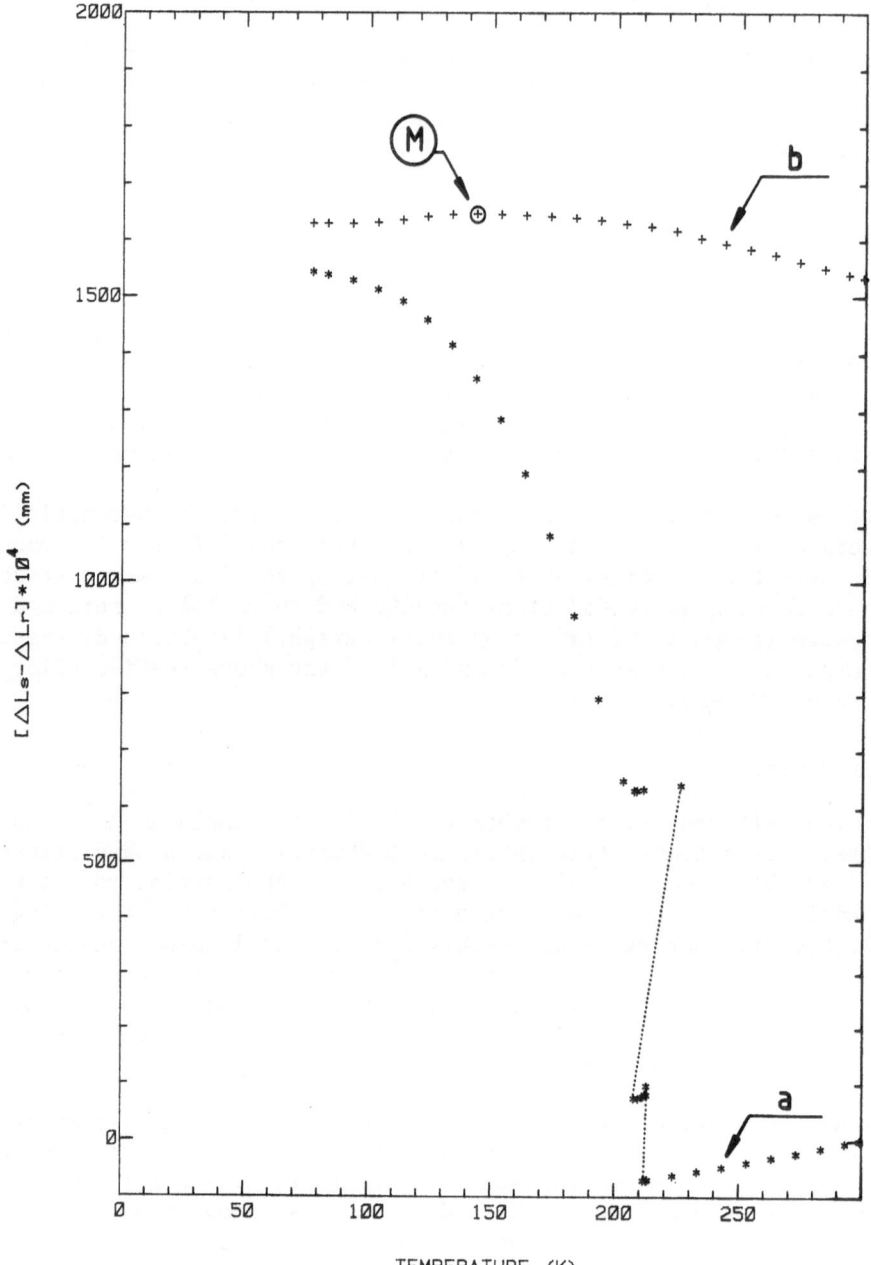

Figure 6. Differential dilatometric behavior of Sample A.

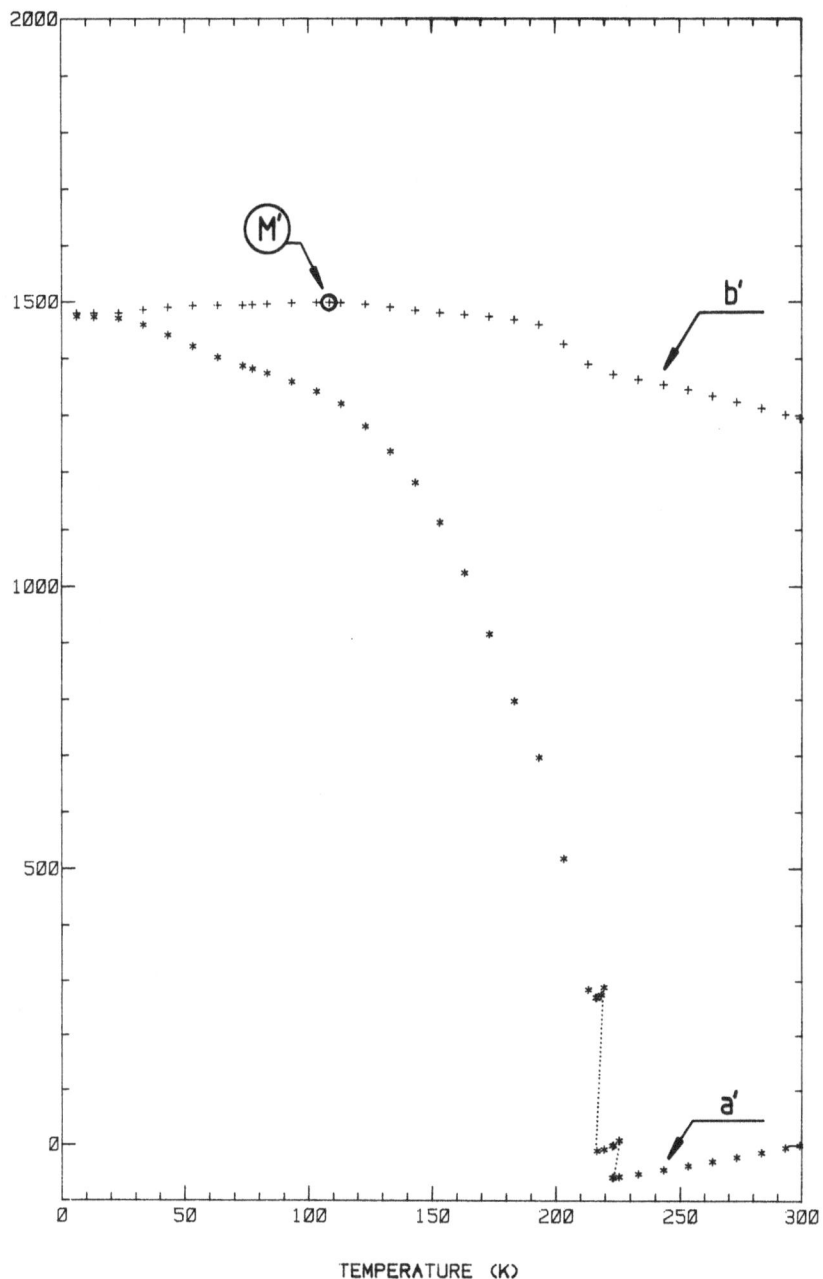

Figure 7. Differential dilatometric behavior of Sample A'.

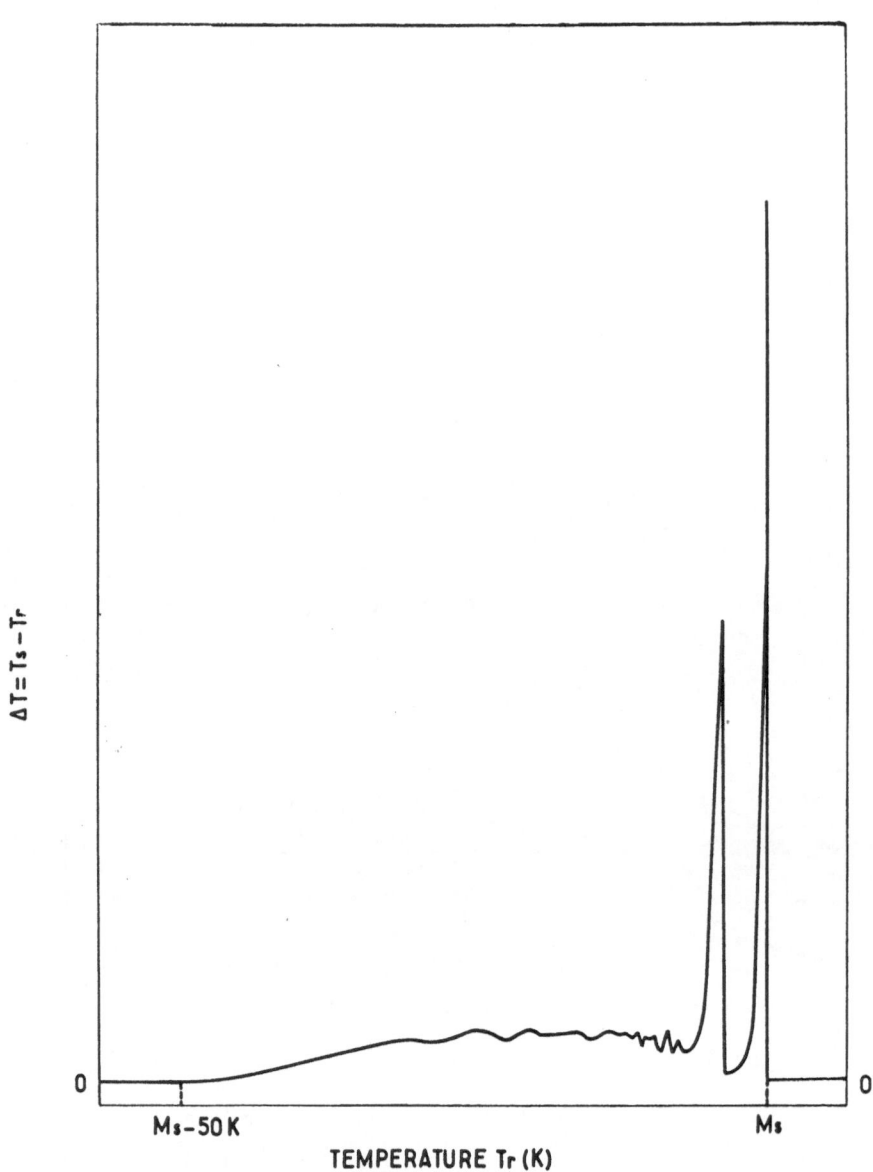

Figure 8.   Typical evolution of the temperature deviation
            $\Delta T = Ts-Tr$ versus the temperature of the reference
            (Tr) for a Fe–Ni–C alloy.

of the reference. At first, the transformation develops by discontinuous detectable successive "explosions" in the Ms to Ms-50 K temperature range. For temperatures below Ms-50 K these explosions are then attenuated. The main maximum of ΔT can correspond to the first or second temperature peak and in some cases the maximum value of ΔT can reach 40 K. This value has to be considered with precaution because of the localized character of the martensitic transformation. This exothermic effect can also be seen in Figures 6 and 7 in the MS-10 K temperature range.

After the first cooling of Sample A to 77 K (Curve a, Figure 6) an isothermal differential expansion $\Delta \ell$ (curve not reported) was observed for 90 hours (limit of experiment). The gap in $\Delta \ell$ at 77 K between Curve a and Curve b corresponds to this isothermal expansion of Sample A at 77 K and is relative to the isothermal transformation of a fraction of retained austenite to martensite.

In former studies[1] an internal friction versus temperature anomaly observed at 120 K in similar Fe-Ni-C alloys was interpreted in terms of a reheat transformation of retained austenite. This effect occurred within the 77 to 170 K temperature range during the first heating following a recently in-situ austenite to martensite transformation from room temperature to 77 K. This interpretation also agreed with dilatometric results obtained with another differential dilatometer.

Curves b and b' in Figures 6 and 7 are in good agreement with the above interpretation. During the first heating of Samples A and A' (Curves b and b', respectively) the irreversible reheat transformation is noticeable. Furthermore, Curve b' in Figure 7 indicates that the reheat transformation can develop from very low temperatures. It is also interesting to note that the maximum M (Sample A) and maximum M' (Sample A') occur at different temperatures indicating that there is a strong influence of the first cool down and subsequent aging on the further development of the reheat transformation of a Fe-Ni-C sample.

Since it is not the purpose of this paper to discuss the results ontained for the Fe-Ni-C samples, but rather to present a description of this new low temperature dilatometer, results concerning only two samples (A and A') of the series already tested are presented here. These results, as well as others, will be discussed in another paper.[2]

REFERENCES

1.   C. Prioul, M. Carrard, J. Plusquellec and P. Azou, Scripta
     Met., 13, 523-526 (1979).
2.   C. A. V. de A. Rodrigues, M. Carrard, J. Plusquellec and P.
     Azou, to be published.

ITES SESSION 3: ITCC/ITES PLENARY SESSION

Session Chairman: I. B. Fieldhouse
                     IIT Research Institute
                     Chicago, IL

THE INTERNATIONAL THERMOPHYSICS CONGRESS - REPORT ON

RECENT DEVELOPMENTS AND FUTURE PLANS

Ared Cezairliyan

Thermophysics Division
National Bureau of Standards
Washington, D.C. 20234

The formation of the International Thermophysics Congress is discussed and some recent developments and future plans are presented.

The objectives of the Congress are to enhance the cooperation among the organizations (conferences and committees) in the thermophysics field and to advance the state of thermophysics. These will be accomplished through:

1. The coordination in the planning of the individual member conferences and the contribution to such conferences through information distribution and participation.

2. The planning of joint conferences and symposia among the member conferences.

3. The cooperation and exchange of information with related national and international conferences.

4. The undertaking of broader activities in the thermophysics area, such as: (a) encouragement of the participation in the coordination of national and international efforts related to reference data and reference materials, (b) the preparation (through task groups) of documents regarding the state of thermophysics and its impact on science and technology, (c) communication with other related fields of science and maintaining awareness among scientific and technical communities in relevant disciplines of the potential of thermophysics, and (d) establishment of closer ties and dialogue between the generators and the users of data in the thermophysics area.

At present, the Congress has the representation of four
Conferences: The Thermophysical Properties Conference of ASME,
The Thermal Conductivity Conference, The Thermal Expansion Conference, and The European Thermophysical Properties Conference. Discussions are underway with other Conferences (in the thermophysics
and related fields) to have representations in the Congress.

ITES SESSION 4: COMPOSITE MATERIALS

Session Chairman: R. K. Kirby
                      National Bureau of Standards
                      Washington, DC

DIFFERENTIAL THERMAL EXPANSION STUDIES OF GRAPHITE

REINFORCED GLASS MATRIX COMPOSITES

F. C. Douglas and K. M. Prewo

United Technologies Research Center

East Hartford, CT 06108

INTRODUCTION

Composites based on graphite fibers have been shown to provide materials with high stiffness-to-weight ratio and sufficient dimensional stability to encourage their use in aircraft and spacecraft applications.[1] Dimensional stability of composite structures becomes more important as the requirements placed on a structure become more stringent with regard to allowable strains, rotations, and surface distortions, such as for mirrors, mirror supports, microwave components, antennas, and similar applications.

Graphite fibers exhibit shrinkage, or a negative expansion, along their length as their temperature is raised. Thus, by layup tailoring of graphite fiber reinforced composites, coefficients of thermal expansion from negative to positive can be obtained, and in particular, a zero coefficient of expansion over a temperature range can be achieved. The problems involved in achieving reproducible zero or near-zero values for composite CTE's are magnified as the temperature range of operation is enlarged, or when the CTE must be particularly uniform and reproducible in addition to being small. Resin matrix composites suffer from problems caused by moisture absorption; these affect the uniformity, reproducibility, and value for designed zero CTE layups. They are also limited in temperature by the properties of the organic matrix component.[2-4] These problems have led to the investigation of graphite fiber reinforced composites employing inorganic glass matrix material.

For the past five years, research and development studies on the fabrication and properties of graphite fiber reinforced glass matrix composites have been pursued at the United Technologies Research Center (UTRC).[5-8] As the processing of these materials has been developed, their mechanical and thermal properties have been measured and used to refine the processing procedures as well as to characterize the material for potential uses. This paper describes the measurement of the thermal expansion characteristics of a graphite/glass composite formed from HMS graphite fiber and "Pyrex" brand Corning Code 7740 glass using a differential thermal expansion apparatus. Axial, transverse (to the fiber direction), and cross-ply laminates were made and measured to examine a variety of thermal expansion characteristics.

THERMAL EXPANSION APPARATUS

A commercially available differential mode dilatometer manufactured by Theta Industries is used at UTRC. The advantages of this type of system when mechanical (i.e. pushrod) dilatometer systems are used has been discussed by Plummer.[9] The differential expansion between a selectable standard and a specimen is transferred to a linear voltage differential transformer (LVDT) through horizontal pushrods attached to the body and to the core of the LVDT which rest against the standard and the sample, respectively. The LVDT body and its core are independently suspended by flexures which eliminate friction from the motion transfer system. The core is mechanically adjustable in position relative to the body of the LVDT so that a zero electrical output can be obtained by mechanical means. Up to 0.020 inches of length difference between the specimen and the standard can be tolerated. The sensitivity of the LVDT is specified at $\pm$ .25 microinches.

The material used in the measuring head section is invar low expansion alloy. Supplementing the use of this alloy, the plate to which the mechanical adjustment pieces are attached is temperature controlled by a 40°C $\pm$ 0.05°C thermostatted circulating water bath, which also provides a thermocouple reference temperature. The head section is enclosed by an aluminum cover with an "O" ring seal for vacuum/inert gas operation. If gas is to be used in the measuring system, it first passes through a coil immersed in the water bath, then through the measuring head volume and finally into the sample region.

The system is supplied with a fused silica sample/reference
holder tube and fused silica pushrods for operation to 800°C, and
with a similar unit made from aluminum oxide for measurement to
1500°C.

The furnace element is a silicon-carbide tube with a reduced-
area central section which provides a 6 inch hot zone. The maximum
temperature difference across the 2 inch sample position is 5°C at
1400°C. Control of the furnace temperature is via a type S
platinum/platinum-10% rhodium thermocouple in a thin aluminum oxide
protection tube placed in contact with the hot zone of the furnace
element. The output from this thermocouple provides the signal for
a closed loop temperature control system. A separate thermocouple,
also type S, is used to record the specimen temperature. This
bare-bead thermocouple is placed in contact with the specimen. Its
cold-junction reference is the 40°C measuring head support plate.
The output of the LVDT measures the differential expansion between
the specimen and the selected reference material. The unit has a
maximum gain of 10,000 which is achieved by the design of the LVDT;
it incorporates a large number of windings instead of using electronic amplification of the LVDT signal.

DATA REDUCTION

The differential expansion signal vs. the specimen temperature
signal is recorded on an analog X-Y recorder using an 11 x 17 inch
format. The data is reduced by digitizing the recorded curve using
a 200 line per inch magnetic wire digitizing board and cursor
(Summagraphics, Inc.). The digitized points are stored in computer
memory, converted to engineering values, corrected for the thermal
expansion of the standard material and for the holder material contribution if the specimen and the standard are not of exactly the
same length, and the result printed as change in length/length from
20°C as a function of the specimen temperature. For visualization
purposes, a graphic display is generated and displayed on a Tek-
tronix CRT terminal from which a permanent copy can be made. The
graphics program has options which enable portions of the curve as
represented by the digitized points to be enlarged to fill the
field of view so that details of the curve can be examined. In
addition, point to point tangents are calculated from the digitized
data to produce an approximation to the derivative of the digitized
curve.

Examples of the data obtained by the above procedure are presented in Figures 1 and 2. Figure 1 illustrates the computer-
generated graphical plot of data generated by digitizing the X-Y
plotter curve produced by the dilatometer when testing a ± 45°
specimen of HMS graphite reinforced 7740 glass matrix composite
between 20°C and 260°C. The computer program has corrected the
measured curve for the effects of the Corning Code 7971 ULE glass
standard and the contribution of the fused silica holder material
for the difference in specimen length. The zero of thermal expansion is set at 20°C. Figure 2 is a plot of the point to point
tangents to the expansion curve calculated from the digitized
curve points. These tangents are plotted vs. the average temperature between the data points. In the case chosen for illustration
the composite is a ± 45°C layup and exhibits a shrinkage of 20 ppm
upon heating from 20°C to about 175°C, and subsequent expansion
with additional heating to 260°C, with a net change of about 9 ppm
of shrinkage. The calculated derivative plot, Figure 2, indicates
a coefficient of expansion varying from -0.3 ppm/°C to +0.2 ppm/°C,
and crosses zero at about 175°C. Under the conditions of running
the instrument for this composite specimen, the uncertainty in the
absolute thermal expansion values is calculated at ± 8 ppm, while
the coefficient of expansion, being the slope of the expansion vs.
temperature curve, has an uncertainty of about ± .01 ppm/°C.

An advantage of a differential thermal expansion system is that
the specimen can be measured relative to a standard of choice where
the thermal conditions during the measurement are the same for both
pieces of material. In many cases the difference in expansion between two materials is specifically the data of interest. This
feature is maintained in the data reduction program in use at the
Research Center. The data presented in Figure 3, with the approximate derivative data in Figure 4, are calculated from the same
digitized curve as the data for Figures 1 and 2, except that no
corrections are applied for the standard material contribution. In
this case, the curves appear nearly identical to those in Figures
1 and 2 since the Corning Code 7971 ULE glass has such a small contribution to the coefficient of expansion in this temperature range.
The derivative curve (Figure 2) is observed to be "noisy", the
value of the derivative, or local tangent, is sensitive to the
exact values of the digitized points used in the computations.
Since these may not be taken in perfectly smooth fashion, an option
for smoothing the data has been incorporated into the data reduction program. This functions by fitting the digitized curve with
a set of cubic splines in the sense of least squares, then

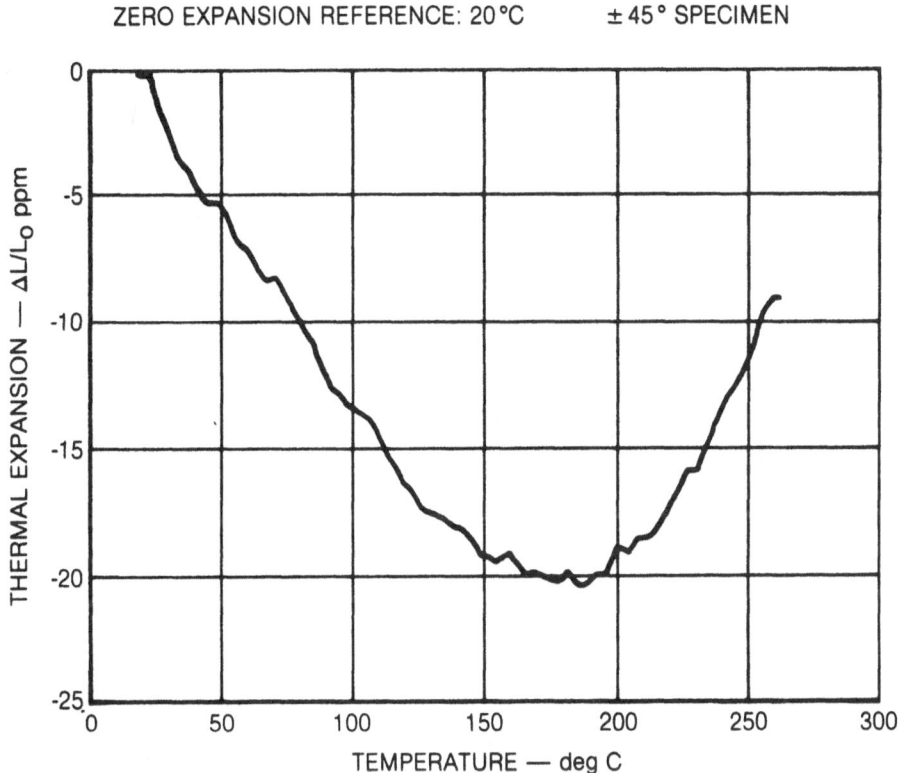

Fig. 1.   Thermal Expansion of Graphite/Glass Composite on
Heating from 20°C to 260°C

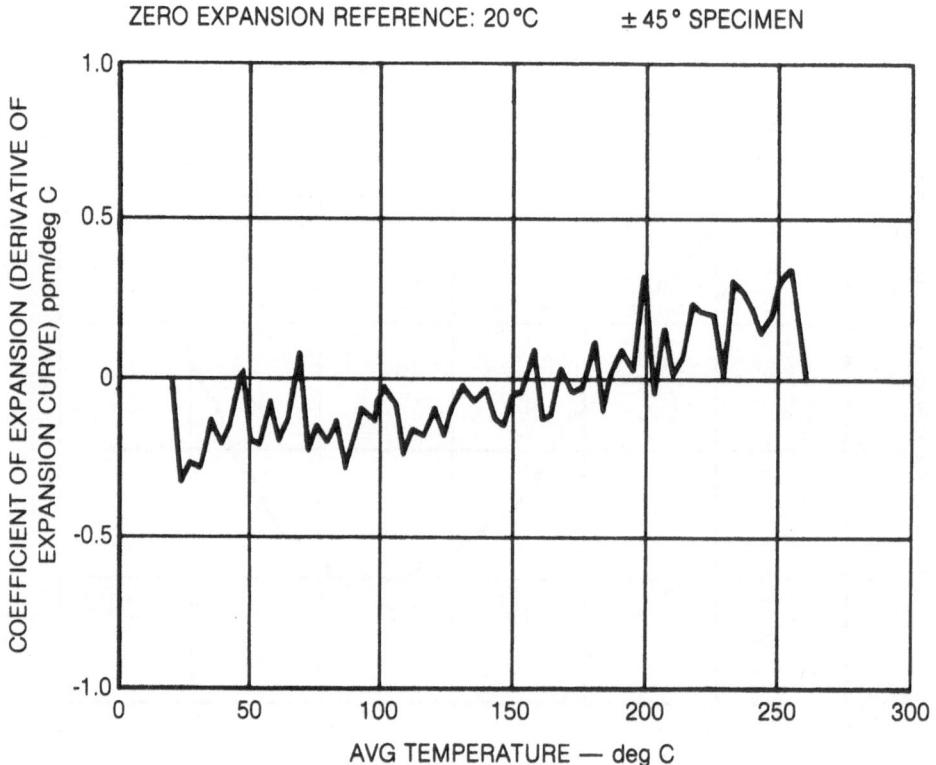

Fig. 2.   Coefficient of Thermal Expansion of Graphite/Glass
          Composite on Heating from 20°C to 260°C

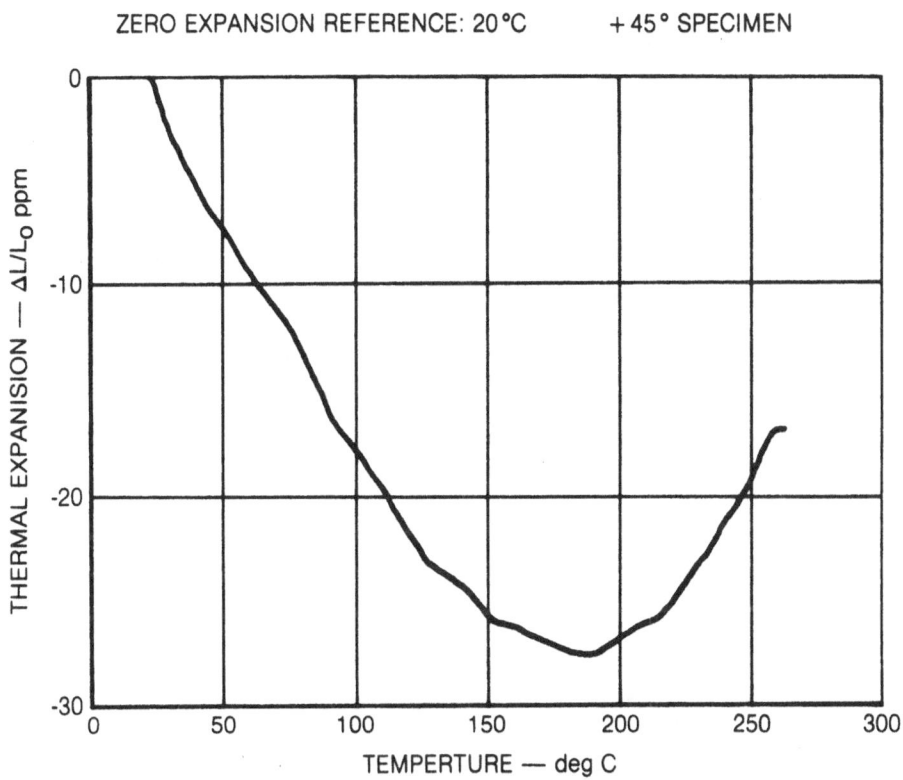

ZERO EXPANSION REFERENCE: 20 °C   + 45° SPECIMEN

Fig. 3.   Thermal Expansion of Graphite/Glass Composite Relative
          to Corning Code 7971 Glass on Heating from 20°C to 260°C

DATA CALCULATED FROM SMOOTHED THERMAL EXPANSION DATA

ZERO EXPANSION REFERENCE: 20°C   +45°C

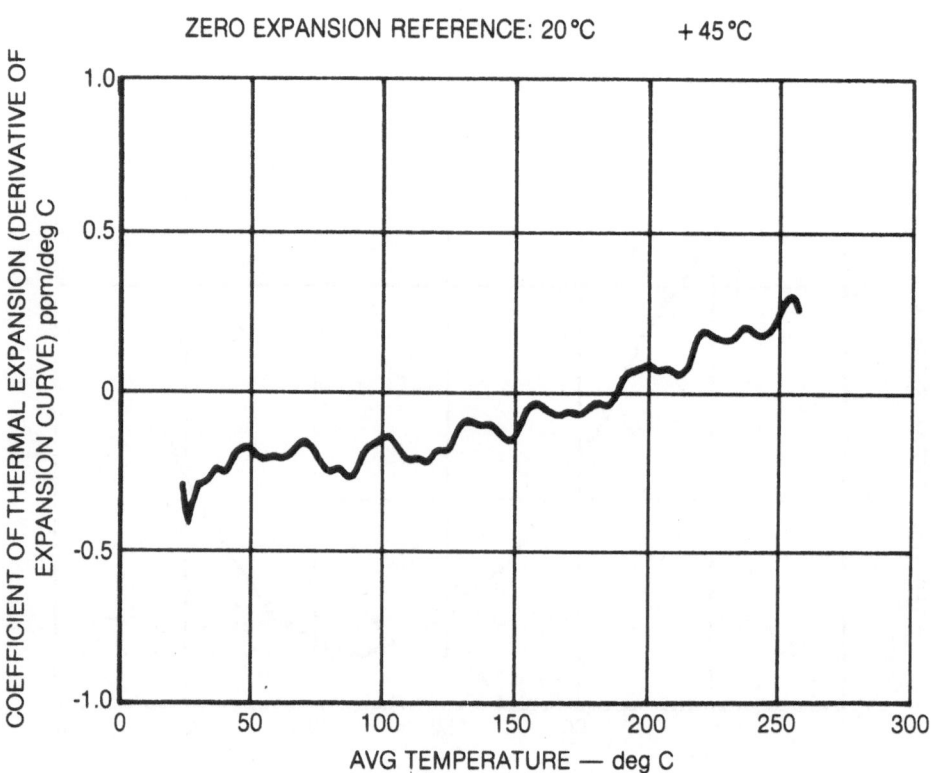

Fig. 4.   Coefficient of Thermal Expansion of Graphite/Glass
Composite Relative to Corning Code 7971 Glass on
Heating from 20°C to 260°C

calculating the local tangents to obtain the derivative curve.
The data of Figure 3 has been subjected to this process, and the
results shown in Figures 3 and 4 are "smoothed" data. The effects
of the smoothing process is most noticeable comparing Figures 2
and 4. An overlay of Figures 2 and 4 shows that the reduction in
the "noise" level by the smoothing routine does not shift the mean
value of the plot.

THERMAL EXPANSION BEHAVIOR OF HMS/7740 GRAPHITE/GLASS COMPOSITES

The composite specimens used for thermal expansion measurements consisted of 0.25 inch x 0.25 inch x 2.0 inch long parallel
sided blocks of graphite fiber reinforced glass material. These
were individually placed in the specimen holder of the dilatometer
in parallel with a Corning Code 7971 silica-titania ultra-low expansion glass sample of nearly the same dimension which served as
the comparative standard. The specimen holder and pushrods were
fused silica. The 7971 standard material had previously been
calibrated by both Corning Glass and the University of Arizona
over a temperature range of -200°C to +600°C.

The operating procedure consisted of thermally stabilizing
the specimens and the instrument by holding at room temperature
for up to 5 hours, followed by a programmed heating rate of 2°C
per minute to the maximum desired temperature of 260°C. There-
after, the specimen was cooled at essentially the same rate by the
use of liquid nitrogen vapor until the minimum temperature of -150°C
was reached. The program then reversed the temperature profile
and returned the specimen to room temperature. The differential
length change was then analyzed by the computer program system
described above. The testing took place in an argon atmosphere to
avoid moisture condensation within the test region.

The axial or 0°, and the transverse or 90°, thermal expansion
data of unidirectionally reinforced HMS/7740 composites is presented
in Figures 5 and 6. These curves represent data obtained on the
second temperature excursion cycle; the first cycle is used to ensure that the instrument and the sample are mechanically "settled"
so that the recorded curve contains no artifacts due to small position changes. At a motion magnification of 10,000, very small
changes produce visible shifts in the data recording.

The axial specimen thermal expansion curve, shown in full in
Figure 7, presents two features of interest. First, the specimen

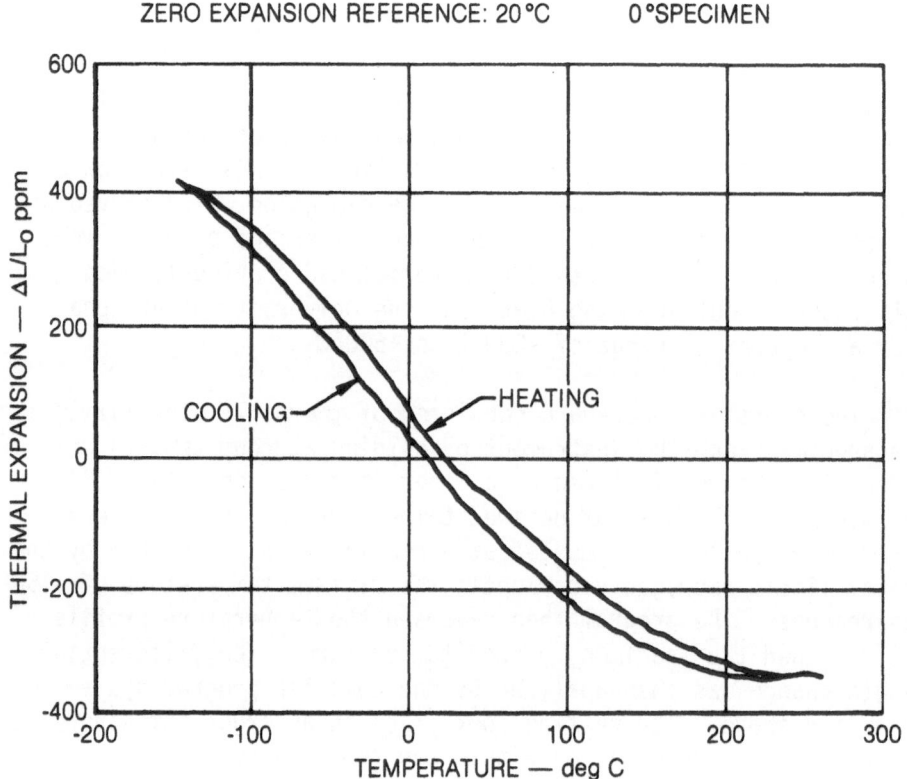

ZERO EXPANSION REFERENCE: 20°C   0°SPECIMEN

Fig. 5.   Absolute Thermal Expansion of Graphite/Glass Composite
on Temperature Cycling Between -140°C to +260°C

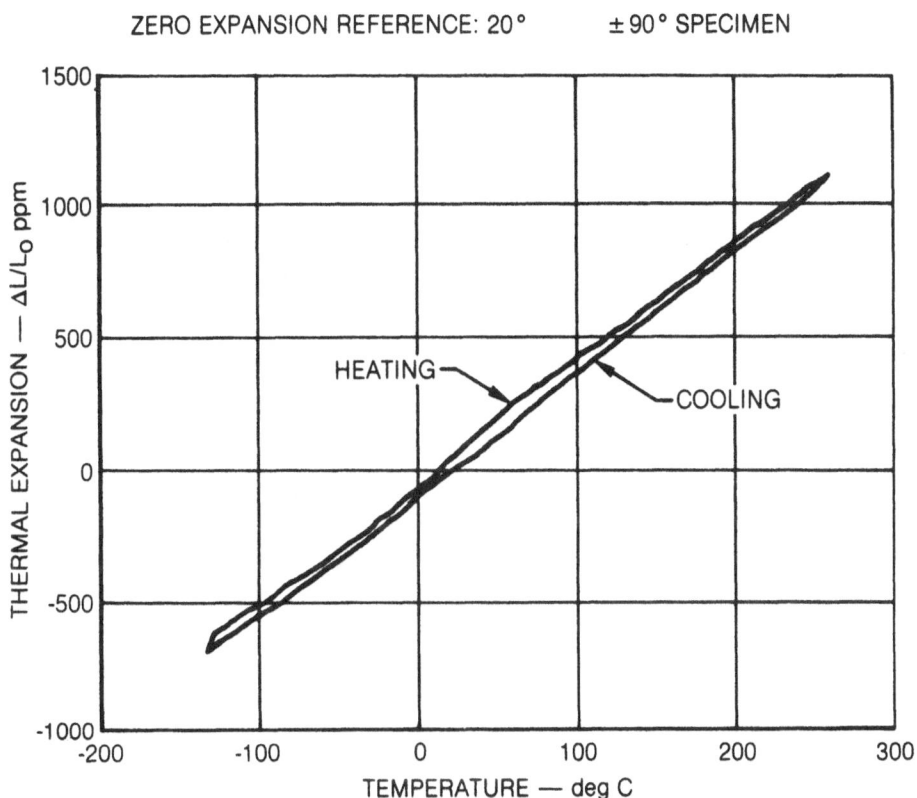

Fig. 6.   Absolute Thermal Expansion of Graphite/Glass Composite
          on Cycling Between -140°C to +260°C

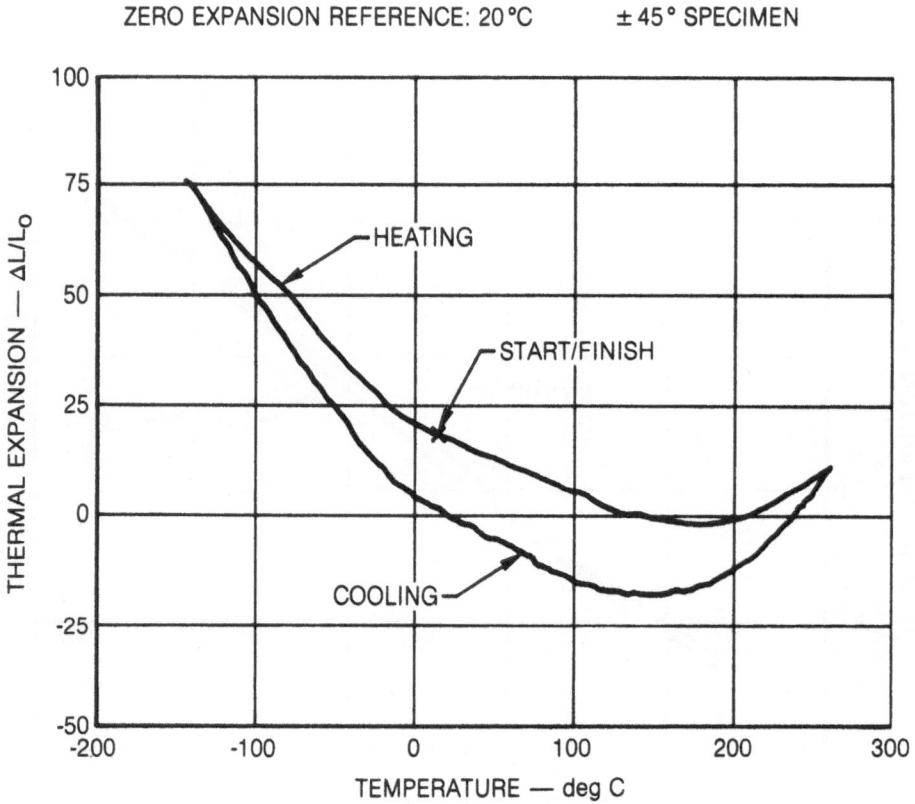

ZERO EXPANSION REFERENCE: 20 °C   ± 45 ° SPECIMEN

Fig. 7.   Absolute Thermal Expansion of Graphite/Glass Composite
          on Cycling Between -140°C and +260°C

contracts as the temperature increases from -150°C to +260°C. The
contraction, as well as its rate, decreases with increasing temperature until at about 240°C, the contraction ends, and the specimen, if taken to higher temperatures, would expand. Thus, the
thermal expansion coefficient passes through zero, and over a
limited temperature range is very small. Over the temperature
range from 20°C to 260°C, the coefficient of thermal expansion
calculated from local tangents to the expansion curve vary from
-0.2 x 10^{-6} in/in °C to zero. A second feature is that the heating
and cooling curves do not appear to coincide; instead, there is
evidence of a hysteresis effect, with the cooling curve of the
specimen indicating more negative strain than the heating curve.
The results of several tests have been examined in an effort to
determine if the hysteresis is actually characteristic of the
specimen, or is due to the operation of the system. Tests in which
the specimen has been cycled from room temperature to 260°C have
shown a hysteresis, but the recorder has returned to the initial
point of the curve, as it did in the test of Figure 7; also, cycling from room temperature to -150°C and return also resulted in
the recorder yielding a hysteresis but returning to its initial
point. Tests have also been run where the program has been stopped
to determine if the rate of heating was affecting the measurements;
in the cases examined, almost no change in the recorded point
occurred. In those cases, such as the present, where the recorder
returns to its original starting point, it is assumed due to some
aspect of the apparatus operation, although the mechanism has not
been identified. It should be noted that the observed magnitude
of the hysteresis is only a few parts per million; this is significant only in cases such as with the present material where the
expansion of the specimen is also only a few parts per million.

The thermal expansion of the transversely oriented specimen,
Figure 6, differs significantly from that of the axially oriented
material. The magnitude of the strain measured is much larger and
is everywhere of positive slope. As in the axial direction measurement, some hysteresis between heating and cooling is observed;
by the above discussion, however, it has not been demonstrated that
the hysteresis accurately represents the specimen behavior. The
expansion curve of the laminated composite in which the fibers are
at ± 45°C to the measured expansion direction is shown in Figure 7.
The scale is an order of magnitude expanded compared to the axial
and transverse measurements. The hysteresis, while apparently
large, is only some 10 to 15 parts per million in the separation
of the heating and cooling curves; the initial and final recorded

Table I

Thermal Expansion Range Over the Temperature
Range from -140°C to +260°C

Orientation	Expansion Range
Uniaxial: Transverse	1800 μinch/inch
Uniaxial: axial	-750 μinch/inch
Cross-ply: ± 45°	90 μinch/inch

points in the cyclic measurement meet within 2 ppm. In summary,
the range of thermal expansion of these layup directions is given
in Table I. It is clear from these measurements that cross-ply
layup of layers in a graphite/glass composite can result in very
small length changes and very small thermal expansion coefficients
over a very useful temperature range.

REFERENCES

1.   E. G. Wolff, "Dimensional Stability of Structural Composites
         for Spacecraft Applications", Metal Progress, June 1979,
         pp 54-63.
2.   W. T. Freemen and M. D. Campbell, "Thermal Expansion Charac-
         teristics of Graphite Reinforced Composite Materials",
         ASTM-STP 497, 1972, p 121.
3.   K. F. Rogers, et al., "The Thermal Expansion of Carbon Rein-
         forced Plastics", Journal of Materials Science, 12, 1977,
         p 718.
4.   R. C. McNamara, "Materials for Large Space Optics, Phase I",
         January 1978, AFML Contract F33615-77-C-5247.
5.   J. F. Bacon and K. M. Prewo, "Research on Graphite Reinforced
         Glass Matrix Composites", NASA Contract Report 145245,
         June 1977.
6.   J. F. Bacon, K. M. Prewo and E. R. Thompson, NASA Contract
         Report 15849, June 1978.
7.   K. M. Prewo and J. F. Bacon, "Glass Matrix Composites", Pro-
         ceedings ICCM-II AIME, April 1978, Toronto, Canada.
8.   K. M. Prewo, J. F. Bacon and D. L. Dicus, "Graphite Fiber Rein-
         forced Glass Matrix Composites for Aerospace Applications",
         SAMPE, May 1, 1979, San Francisco, California.
9.   W. A. Plummer, "Differential Dilatometry: A Powerful Tool",
         AIP Conf. Proc. #17, pp 147-158, "Thermal Expansion-1973",
         ed. R. E. Taylor and G. L. Denman.

ISOTROPIC ZERO CTE MATERIALS

G. F. Hawkins and E. G. Wolff

Aerospace Corporation
2350 East El Segundo Blvd.
El Segundo, California 90245

INTRODUCTION

Stringent dimensional stability requirements are called for in present and future aerospace systems, components and structures. The availability of a zero coefficient of thermal expansion (CTE) material would help to solve numerous design and performance problems. At present the best materials are ULE, Zerodur, fused SiO_2, Super Invar and some pseudo-isotropic graphite epoxy composites. All of these exhibit thermal strains ($\Delta L/L$) in excess of 20-50 microinches per inch when cooled, for example, from room temperature to -250°C. Since immediate applications for a zero CTE material include mirror substrates, metering rods and optics supports, an optimum material should also be isotropic, dimensionally stable with time (under constant temperature and low stress levels) and exhibit as high a thermal conductivity as possible.

Zero CTE does exist, but only in small temperature regions where there is a CTE sign reversal (e.g. ULE near R.T., and SiO_2 near -100°C). Optics and sensors are frequently required to operate near liquid nitrogen temperatures and then both the average CTE to the operating temperature and dimensional changes at the operating temperature become problems. In addition, the thermal cycling required during testing and the need for minimal outgassing must also be considered.

An approach to the development of isotropic zero CTE materials is described wherein powders from positive and negative CTE materials are hot pressed at optimum volume fractions. Table I lists some candidate constituents with CTE values generally of $0 \pm 1 \times 10^{-6}$ C^{-1} in the $\pm 250^{\circ}$C range. $^{1-6}$ The theory, fabrication, test methods, and constituent characterization are described in the following sections.

THEORY

Atomistic adjustments to a homogeneous material (including compositional, thermomechanical or electromagnetic treatments) are likely to yield a zero CTE only over small temperature regions. Consequently, a composite is considered to provide the best approach to zero CTE over $\pm 250^{\circ}$C. The following theoretical approach does not dictate the fabrication method, which should be selected after the composition has been chosen.

The first step is to characterize the expansion behavior of isotropic constituents, e.g. powders, through use of polynomials.

$$\left(10^6\right) \frac{\Delta L}{L} = d_o + d_1 T + d_2 T^2 + d_3 T^3 + d_4 T^4 -----d_n T^n \qquad (1)$$

$$\left(10^6\right) CTE = d_1 + 2d_2 T + \frac{3d_3}{2} T^2 + \frac{4d_4}{3} T^3 --- \frac{nd_n T^{n-1}}{n-1} \qquad (2)$$

Several computer programs have been developed to provide the coefficients. These include a BASIC program to give third order coefficients. The HP67-ST1-14A1 and ST1-14A2 polynomial approximation program is useful for 4th order (5 coefficients) provided the temperatures are entered at regular intervals. An NFIT program for Nth order polynomials has been programmed in FORTRAN for use with the Control Data Corporation 7600 Computer. Plots are made automatically to assure the curve fits the data points to $\leq 1 \times 10^{-6}$ in $\Delta L/L$. Table II lists some typical coefficients obtained from recently reported data. As will be shown later, these numbers must be modified to correspond to the particular fabrication sequence employed. Figure 1, for example, illustrates the effects of heat treatment and impurity concentrations on the expansion behavior of nominally pure SiO$_2$[7].

Table I - Candidate Constituents for
Isotropic Zero CTE Composite Materials

A. Materials with Positive CTE's of $<5 \times 10^{-6}$ C^{-1} in range
    150-450K.

Grade A graphite	Al-Ti-O systems
AlN	Al-Li-Mg-Si-O
BN	Si-Zr-O
SiC	Invar (Fe-35 to 36 Ni)
Si_3N_4	Nb-Hf-O
BP	
RVC Foam No. 45	
UCC grade VT-6	
Corning glass No. 9618	

B. Materials with Negative CTE's (and systems wherein
    negative CTE's have been reported).

$NiO + 20\%$ CaF_2	Li-Ta-O
$NiO + 12\%$ $CaF_2 + 38\%$ C	Li-Al-Ti-Si-O
$Al(OH_3)$ (Gibbsite)	Li-Al-Si-O
Nb_2O_5	Al-Ti-O
Hercuvit (PPG) glass Ceramic	W-Zr-O
Pycobond Gr Pitch	Ta-V-O
$KAl\,Si_3O_8$	Hf-Ti-O
	Ta-W-O
	Hf-Ta-W-O
	Mg-Al-Li-Si-O
	Pb-Ti-Zr-O
	Ge-Al-Li-Si-O
	Zn-Al-Si-O

C. Materials with both low positive and negative CTE's in
    range 150-450K.

Si	Li-Ta-O
Nb_2O_5	Pb-Ti-O
SiO_2	Al-Li-Si-Mg-O
V_4O_9	Al-Li-Si-Ti-O
Pyroceram 9606 (Corning)	Ti-Si-O
Super Invar (Fe-Ni-Co)	
ULE ($SiO_2 + 7\%$ TiO_2, Corning)	
Zerodur (Schott)	
CER-VIT (Owens Illinois)	
ZnO	

Table II.   Expansion Coefficients

Material	T_{min} °C	T_{max} °C	d_0	d_1	d_2	d_3	d_4	Ref. for Data
Invar 36 (ann).	-73	227	-3.278	0.151	6.543 E-4	-1.3559 E-6		This work
Invar LR35	-203	-98	-35.76	-2.15	-2.77 E-2	-6.886 E-5		This work
TiO$_2$	-173	227	-151.53	7.508	3.598 E-3	-7.836 E-6		(3)
SiC	-123	227	-66.3496	3.162	8.21 E-3	-2.195 E-5		(3)
WC	20	227	-89.35	4.616	-7.8958 E-3	2.466 E-5		(3)
Si	-253	227	-0.4626	2.212 E-2	5.13x10^{-5}	-4.499 E-8		(3)
Mg	-268	227	-507.35	24.7676	3.024 E-2	-1.328 E-5		(3)
Al	-93	327	-4.17	0.204	1.780 E-4	-7.690 E-8		(3)
SiO$_2$-739	-193	107	9.32	0.43	1.4624 E-3	-5.228 E-6	-2.251 E-10	NBS SRM 739
SiO$_2$-7940	-200	200	45.42	0.24	-1.290 E-3	2.660 E-6		Corning
Zerodur	0	200	-1.61	-0.10057	-1.440 E-3	4.590 E-6		U. Ariz.
Cervit	-250	+250	21.196	-0.455	9.830 E-4	2.604 E-6		U. Ariz.
ULE	-150	150	0.643	-0.047	7.850 E-4	-1.260 E-6		U. Ariz.
Pycobond Gr/ep	-124	24	11.83	-0.483	-2.439 E-4	-2.355 E-6		(1)
ThF$_4$	25	200	153.04	-8.06	0.09	-5.03 E-4	1.042 E-6	(5)
Invar 36 (ann).	-73	227	-3.278	0.151	6.543 E-4	-1.3559 E-6		This work
Invar LR 35	-203	-98	-35.76	-2.15	-2.77 E-2	-6.886 E-5		This work
TiO$_2$	-173	227	-151.53	7.508	3.598 E-3	-7.836 E-6		(3)
SiC	-123	227	-66.3496	3.162	8.21 E-3	-2.195 F-5		(3)
WC	20	227	-89.35	4.616	-7.8953 E-3	2.466 E-5		(3)
Si	-253	227	-0.4626	2.212 E-2				(3)

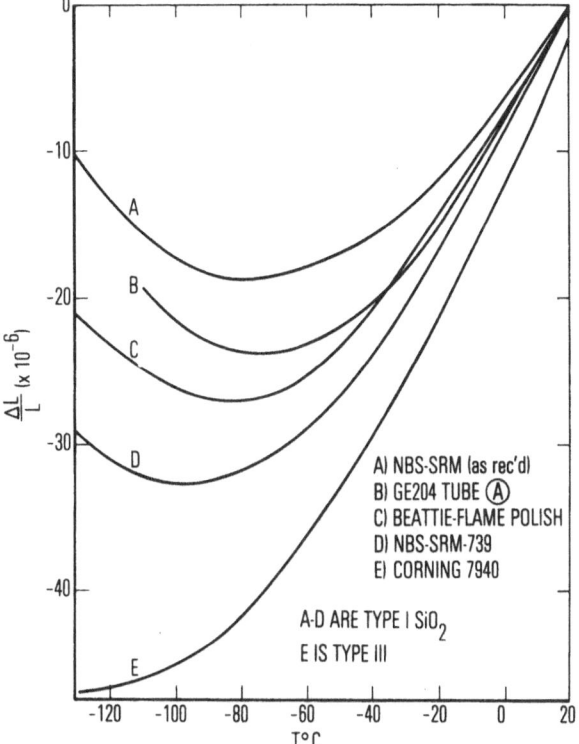

Fig. 1   Low temperature expansion of some pure fused silicas

Another FORTRAN program takes N constituents (each described by any number of coefficients) and computes

$$\epsilon_c = \frac{\sum\limits_{i}^{N} V_i \epsilon_i(T)\ (E\ or\ K)_i}{\sum\limits_{i}^{N} V_i\ (E\ or\ K)_i} \tag{3}$$

where V_i is the volume fraction of each constituent, $\epsilon = \Delta L/L$, and E and K represent the elastic and bulk modulus. This is the Turner equation for ϵ_c or CTE of a composite. The computer then calculates

$$\bar{\epsilon}_c = \frac{1}{T_2 - T_1} \int_{T_1}^{T_2} \epsilon_c (V, E, T) \, dT \tag{4}$$

and remembering $\sum_{i}^{N} V_i = 1$, differentiates

$$\frac{\partial \bar{\epsilon}_c}{\partial V_1} = 0, \quad \frac{\partial \bar{\epsilon}_c}{\partial V_2} = 0, \quad \frac{\partial \bar{\epsilon}_c}{\partial V_n} = 0 \tag{5}$$

This determines the values of V_1, V_2 --- V_N to give the minimum average expansion or contraction of the composite between the chosen temperatures T_1 and T_2. The computer then plots out $\bar{\epsilon}_c = f(V_1, V_2 --- V_N)$ to determine that there is a non-zero solution, that is, whether the constituent is capable of lowering $\bar{\epsilon}_c$ when $V_i > 0$. The computer next plots $\bar{\epsilon}_c$ vs. T to illustrate the range of maximum and minimum deviations of $\bar{\epsilon}_c$ values. Thus a criterion for a zero CTE is not only a minimum average strain $(\bar{\epsilon}_c)$ over a chosen temperature interval but also a maximum absolute deviation of $\bar{\epsilon}_c$ from zero over this range.

It is necessary to know the validity of the Turner equation (Eq. 3). There is reason to believe that composite strains based on particulate constituents may be better described by the Kerner equation [8, 9].

$$\epsilon_c = V_1 \epsilon_1 + V_2 \epsilon_2 - (\epsilon_1 - \epsilon_2) V_1 V_2 \theta \tag{6}$$

where

$$\theta = \frac{(1/K_1) - (1/K_2)}{\dfrac{V_2}{K_1} + \dfrac{V_1}{K_2} + \dfrac{3}{4G_1}} \tag{7}$$

and G is the shear modulus.

The final theoretical task involves combining the other material requirements into an analytical form. For example, use of a generalized figure of merit could be made:

$$\beta(V, T) = \frac{\epsilon^a \rho^b \sigma_i^c}{d^d E^e} \tag{8}$$

where k is the thermal conductivity, ρ the density and σ_i the interfacial stresses between composite constituents. The co-efficients a, b,—— e represent weighting factors.

The residual stresses might be described according to

$$\sigma_i = \frac{(CTE_1 - CTE_2)(T_{fab} - T_{op})}{\dfrac{1 + \nu_1}{2E_1} \quad \dfrac{1 - 2\nu_2}{E_2}} \tag{9}$$

where constituents 1 and 2 represents the extremes (out of N constituents) in CTE values and ν's are the corresponding Poisson's ratios. Preliminary calculations suggest that an Invar/ULE composite would generate stresses in the ULE of ~1/10 MYS and for Invar about 1/4 MYS (microyield strength). The computer would then calculate

$$\beta_c = \frac{1}{T_1 - T_2} \int_{T_1}^{T_2} \beta(V, T) \, dT \tag{10}$$

and $\quad \dfrac{\partial \beta_c}{\partial V_1} = 0, \quad \dfrac{\partial \beta_c}{\partial V_2} = 0 \quad$ etc. $\tag{11}$

The optimum V_i values can then be compared to the values selected on the basis of CTE only. This would indicate whether a near zero CTE material could also withstand thermal cycling without microcracking and could show adequate response to thermal excursions in general.

FABRICATION STUDIES

Vacuum hot pressing of constituent powders was chosen as the fabrication method because of the high densities achievable at moderate temperatures, minimal chemical interactions be-tween constituents, and the convenient shape for CTE measure-ment. Higher fabrication temperatures, as with sintering,[10] or casting are likely to result in excessive grain growth and re-sidual stresses on cooling. ATJ graphite dies and plungers, graphite wool insulation and BN wash coatings to minimize sample - die interactions were employed. Standard hot pressing conditions were chosen as 1000 C, 5000 to 7000 psi pressure, and 1 hour at temperatures followed by furnace cooling. No binders or sintering aids were used. Super Invar, Invar, Nb_2O_5 and SiO_2

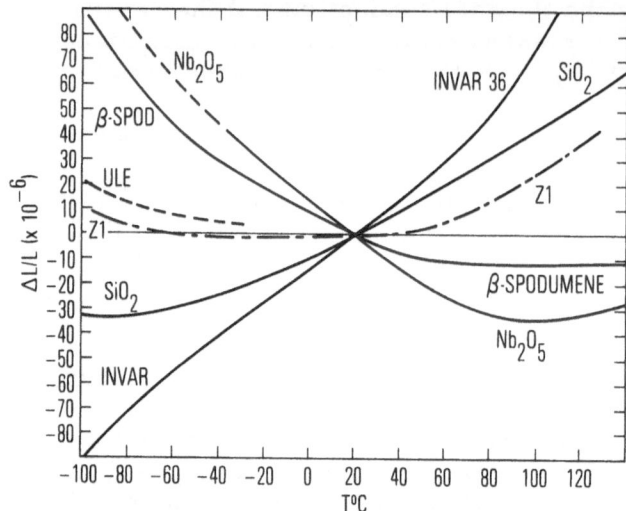

Fig. 2 Calculated curve (Z1) based on 25 v/o of each of the other
4 materials from literature data

powders were successfully hot pressed to over 95% theoretical
density. In order to test the feasibility of consolidating diverse
constituents in the same manner, 25 volume percent each of Invar
(36% Ni), Nb_2O_5, SiO_2 and β-spodumene powders were mixed
and hot pressed. (The data of Table II suggested that such a
mixture would have favorable CTE behavior, as shown in Figure
2.) Figure 3 shows the microstructure in the pressing direction
and little porosity and good interfacial adhesion was found. No
attempt had been made to minimize the grain size in this case.
Nevertheless, the 1/2 -in. diameter by 1-in. long sample could
be quenched indefinitely from red heat into LN_2 without visual
deterioration - an indication of excellent thermal shock properties.
The measured CTE, however, was found to be in the 1 - 2 x 10^{-6}
C^{-1} range. Reasons for this are described in the results section.

Fig. 3 Hot pressed mixture (Z1) of 25% each Invar, β-Spodumene,
Nb$_2$O$_5$ and SiO$_2$ Powders   100X

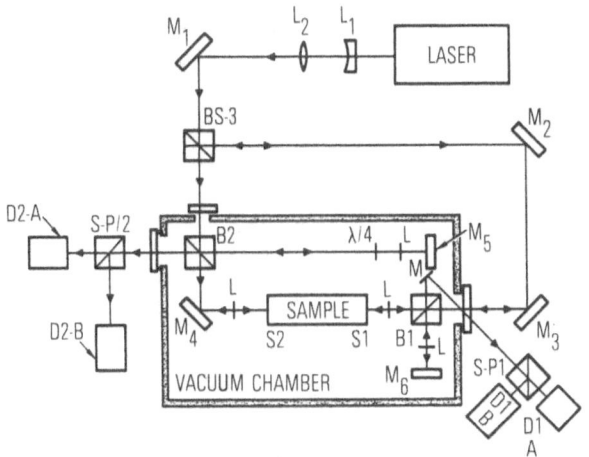

Fig. 4 Double Michelson Laser Interferometer

CTE MEASUREMENT

The sample ends were polished and vapor coated with Cr, Au and/or Al. The expansion characteristics were measured in a vacuum by reflecting 200 μW He-Ne laser beams off the ends as part of a double Michelson interferometer system,[11] also illustrated by Figure 4. Automatic real time recording of the end face motions is accomplished by circularly polarizing the light beams and analyzing the fringe patterns with S-P beam splitters and four photodetectors. The amplified analog signals are recorded on strip charts. Vacuum pump vibration presently limits the resolution to $\lambda/24$ or about 1.03×10^{-6} inch. For a 1" sample and a 500 C temperature excursion, this represents a CTE resolution of about 2×10^{-9} C^{-1}. At these low CTE values, special precautions are required. For example, thermal gradients in the sample must be kept low enough to avoid ΔT errors.[7] The sample position must remain exceptionally stable during temperature excursions. This may require the use of an identical material as the test sample to support the latter. The time stability of the interferometer becomes critical. In the present system, it was found that over six hours after a vacuum of $\leq 10^{-4}$ TORR had been reached was required to achieve a drift rate of $\leq \lambda/120$ per minute (Fig. 5). A correction for the thickness of the reflective coating will also be ultimately required.

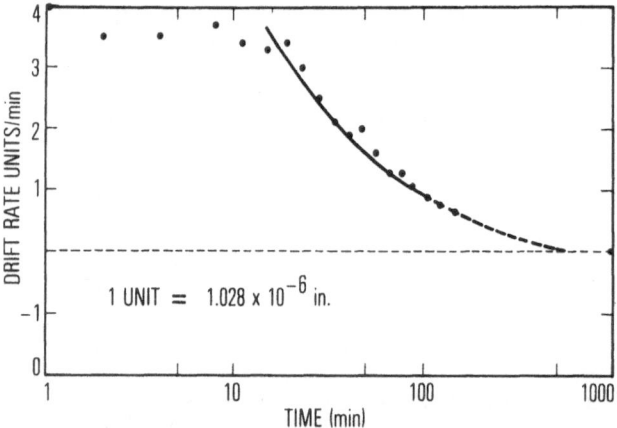

Fig. 5 Drift Experiment

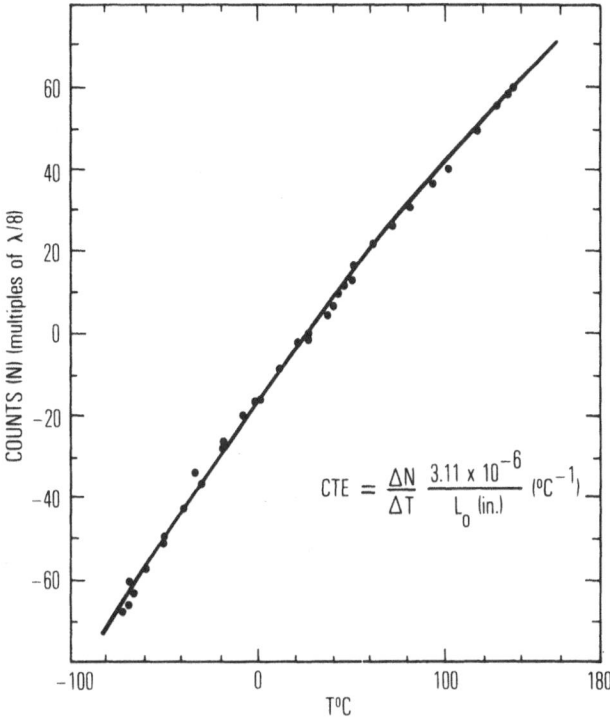

Fig. 6   Thermal expansion of hot pressed Invar L_o = 0.82 in.

PRELIMINARY RESULTS

Hot pressed Invar (36% Ni) gave expected CTE results (Fig. 6).
This sample had a density of 96.3% theoretical and the CTE is
seen to be near 2×10^{-6} C^{-1}. This is characteristic of well
annealed, furnace cooled (from over 800°C), Invar with $\geq 0.07\%$ C.

The search for a suitable negative CTE material has proven
more difficult. Values as low as -2×10^{-6} C^{-1} have been reported
for Nb_2O_5 in the vicinity of room temperature[3, 10]; hence, this
material was chosen for initial investigation. Powder 99.9% pure
was hot pressed to a density of $4.56 \pm .02$ g/cc which is essentially
100% dense. Figure 7 shows a fine grained material with very
little porosity. The CTE data are presented in Figure 8. The
first (cooling) cycle showed two small negative CTE regions -
one just below R.T. and the other just below -70 C, resulting in
a net expansion on reheating. The next (heating) cycle also showed
a net expansion on returning to R.T., but the third cycle to 100°C
showed no hysteresis. Half of the original sample was heated in
air at 1000°C for 8 hours which restored the original white color
of the powder. This sample was heated twice to 100 C for CTE
measurements (Fig. 8).

Fig. 7 Microstructure of Nb_2O_5 hot pressed at $1000^\circ C$, 1000 psi
        1 hour, HF etch   1000X

        Figure 9 shows again the dependency of the CTE on the
fabrication method. The cold pressed and sintered, porous
(δ = 1.36 g/cc) β-spodumene showed a slightly negative CTE
and much hysteresis on its first heating and cooling cycle.
(The material is called Super Beta-K, Centerflex Ceramics
Corporation, Hawthorne, New Jersey.) The same material,
screened to -325 mesh, when hot pressed was grey in color and
had a density of 1.96 g/cc. The CTE is about 0.8 to 1.1 x 10^{-6}
C^{-1} between -80 and $110^\circ C$.

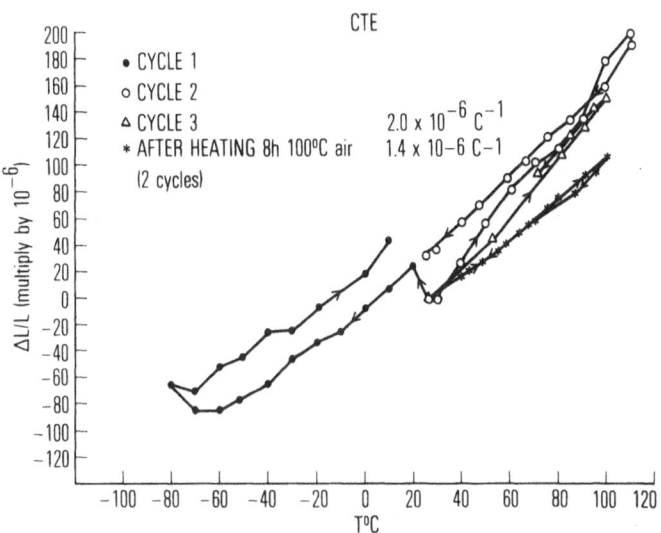

Fig. 8   Thermal expansion of hot pressed Nb_2O_5

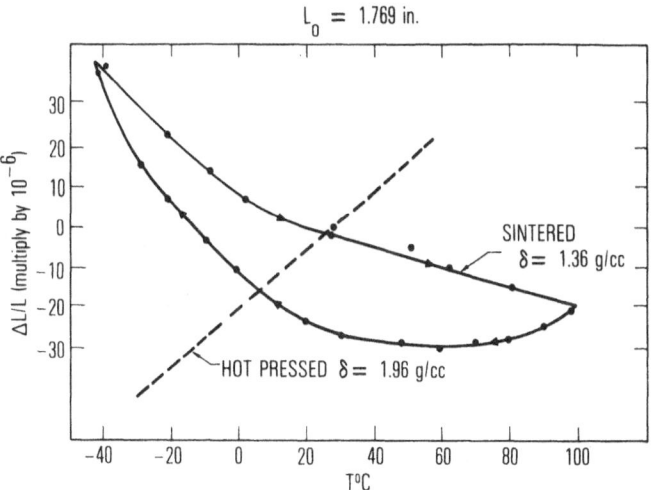

Fig. 9   Thermal expansion of β-Spodumene

DISCUSSION

The Nb_2O_5 data are consistent with the results of Manning et al[4] and Douglass. [12]   Sintering of cold pressed ceramics generally results in a greater grain size than hot pressing (e.g. above ~ 3 μm for Nb_2O_5).   This accentuates the effects of crystalline anisotropy on subsequent cooling. Microcracking relieves the stresses between adjacent grains causing an apparent expansion and a negative apparent CTE. This would appear to be analogous to the cracking expansion found in graphite fiber reinforced composites.[13] The present data suggest that a similar effect occurs for hot pressed Nb_2O_5, but at lower temperatures than normally found with sintered material. Frequency spectrum analysis of the interferometric signals below - 70°C would be instructive. [13]

If microcracking effects are used to obtain apparent negative CTE's, then their modification with additional constituents in a composite must be determined. For Nb_2O_5, at least, it does not seem possible to avoid microcracking if a negative CTE is desired. The expansion due to microcracking would be difficult to predict because of its inferred dependence on grain size, crystal structure and porosity. Even if a single negative CTE material can be fabricated and tested free from microcracks, the changes in interfacial stresses when it is incorporated into a composite must still be considered. Thermal cycling may stabilize a given CTE but the use of that value in a multi-constituent property equation remains questionable.

Douglass[12] showed that stoichiometric Nb_2O_5 has a slightly lower CTE than $Nb_2O_{4.978}$. The present data substantiate this observation. Chemical analysis of the super Beta-K material indicated that it contained 45 wt % Si, with Li, Al, Pb and Na with amount range 0.1 to 1%. This suggests a deviation from the range of β-spodumenes.   The CTE of β spodumene-silica solutions ($Li_2O \cdot Al_2O_3 \cdot nSiO_2$) depends strongly on the value of n. [6] The volume change from R.T. to 100°C changes from positive to negative as n goes from 6 to 7. If microcracking can be prevented, suitably negative CTE materials might be prepared in this system.

CONCLUSIONS

An approach to the development of isotropic CTE materials has been outlined. It is based on an optimized composition fabricated by hot pressing to a fine grained highly dense material. Preliminary studies have defined the theoretical and measurement requirements and have indicated the importance of fabrication parameters. Stoichiometry, porosity, grain size, microcracking and heat treatments are interrelated and critical to the CTE values. Techniques such as opto-acoustic analysis[13] should prove useful in monitoring microcracking in ceramics as well as other composites. Microcracking as a means of achieving a negative CTE is considered to be undesirable because of lack of predictability and reproducibility when that material is used as a constituent in a composite. Additional factors which are expected to become relevant in this study include impurities,[7] crystallinity, phase transitions, stress relaxation or time effects, plastic flow, and end or edge effects.

ACKNOWLEDGMENTS

This work was sponsored by the U. S. Air Force and the Defense Advance Research Projects Agency under Contract F04701-78C-0079.

REFERENCES

1.   R. C. McNamara et alia, "Materials for Large Space Optics, Phase I Final Technical Report," AFML-TR-78-12 (1978)

2.   S. R. Skaggs, "Zero and Low Coefficient of Thermal Expansion Polycrystalline Oxides," Los Alamos Report No. LA-UR-77-1307 and presented at Int. Colloquium on Refractory Oxides, Odeillo, France (1977)

3.   Y. S. Touloukian, R. K. Kirby, R. E. Taylor and T. Y. R. Lee, Thermophysical Properties of Matter-TPRC Data Series, Volumes 12 and 13, (Thermal Expansion-Metallic Elements and Alloys, and Non metallic Solids/IFI/Plenum Press (1977)

4.   W. R. Manning, O. Hunter, F. W. Calderwood and D. W. Stacy, "Thermal Expansion of Nb_2O_5." J. Am. Ceram. Soc., 55, pp. 342-347 (1972)

5.   L. G. Van Uitert et alia, "Physical Properties of Thorium
      Fluoride," Met. Res. Bull, Vol. 11, pp 669-672 (1976)

6.   W. Ostertag, G. R. Fisher and J. P. Williams, "Thermal
      Expansion of Synthetic β-Spodumene and β-Spodumene-
      Silica Solid Solutions," J. Am. Ceram. Soc., 51, (11)

7.   E. G. Wolff and S. A. Eselun, "Thermal Expansion of a
      Fused Quartz Tube in a Dimensional Stability Test Facility,"
      Rev. Sci. Instr., 50, (4), 502-506 (1979)

8.   L. Holliday and J. Robinson, "The Thermal Expansion of
      Composites Based on Polymers," J. Mat. Sci., 8,
      301-311 (1973)

9.   E. H. Kerner, "The Elastic and Thermal Elastic Properties
      of Composites Media, " Proc. Roy. Soc. London, 69 B(8),
      pp 808-813 (1956)

10.   G. D. Kidwell, J. H. Richardson and G. M. Wolten, "Thermal
      Expansion Measurements in the Nb_2O_5 - Ta_2O_5 System,"
      Aerospace Report TR-0172 (2250-40)-1, 15 December 1971

11.   E. G. Wolff and S. A. Eselun, "Double Michelson Interfer-
      ometer for Contactless Thermal Expansion Measurements,"
      SAMSO-TR-78-136 (1978); also pp 204-208 in "Interfer-
      ometry," SPIE Proceedings, Vol. 192 (1979)

12.   D. L. Douglass, "The Thermal Expansion of Niobium
      Pentoxide and Its Effect on the Spalling of Niobium
      Oxidation Films," J. Less-Common Metals, 5, pp.
      151-157 (1963)

13.   S. A. Eselun, H. Neubert and E. G. Wolff, "Microcracking
      Effects on Dimensional Stability," 24th SAMPE Symposium;
      San Francisco, CA, May 8-10, 1979, pp 1299-1309

ITES SESSION 5: THEORY AND CORRELATIONS

Session Chairman: P. S. Gaal
                     Anter Laboratories
                     Pittsburgh, PA

ROLE OF THERMAL EXPANSION IN THERMAL STRESS RESISTANCE OF SEMI-ABSORBING BRITTLE MATERIALS SUBJECTED TO SEVERE THERMAL RADIATION

J. R. Thomas, Jr., J. P. Singh, and D. P. H. Hasselman

Departments of Mechanical and Materials Engineering
Virginia Polytechnic Institute and State University
Blacksburg, Virginia 24061

ABSTRACT

An analysis is presented of the thermal stresses resulting from "thermal trapping" in semi-absorbing brittle ceramic materials in the form of a thin flat plate subjected to intense thermal radiation. Solutions for the thermal stresses are presented for symmetric and asymmetric radiation heating and convective cooling with limiting values of the heat transfer coefficient, $h = 0$ and ∞. For $h = \infty$, the stresses are identically equal to zero for values of the optical thickness μa of 0 and ∞, and reach their maximum value at $\mu a \simeq 1.3$ and 2.0 for symmetric and asymmetric heating respectively. For $h = 0$, the magnitude of thermal stress increases with increasing optical thickness.

Expressions are derived for the maximum radiation heat flux to which the plate can be subjected without failure, in terms of the pertinent material properties. These properties are combined in "figures-of-merit" for the selection of material with the optimum thermal stress resistance, appropriate for given heating and cooling conditions. These figures-of-merit, which may contain as many as seven material properties, indicate the high thermal stress resistance requires low values of the coefficient of thermal expansion, Young's modulus, Poisson's ratio, emissivity, and the absorption coefficient in combination with high values of tensile strength and thermal conductivity.

121

INTRODUCTION

Non-uniform thermal expansions in materials or structures subjected to non-linear temperature distributions result in thermal
stresses. The magnitude of these thermal stresses can be sufficiently high to result in catastrophic failure. This problem is particularly severe for brittle materials which do not permit thermal
stress relief by plastic flow. Thermal stress analysis is (or should
be) an essential element in the design of engineering structures or
components operating under non-uniform temperature conditions.

In practice, the problem of thermal stress failure can be
solved by a structural re-design. An alternative solution is to
select a material with the optimum combination of properties which
keep the magnitude of the thermal stresses well below the failure
stress. The magnitude of thermal stress is not only a function of
the coefficient of thermal expansion, but also depends on the elastic properties, the thermal conductivity and diffusivity and many
other properties depending on the nature of heat transfer, performance criteria and mechanisms of failure [1]. The choice of the
optimum material to avoid thermal stress failure for a given thermal environment can be based on "thermal stress resistance parameters" or "figures-of-merit." The two best known thermal stress
resistance parameters are [1,2]:

$$R = S_t(1 - \nu)/\alpha E$$

and
$$R' = S_t(1 - \nu)k/\alpha E,$$

in which S_t is the tensile strength, ν is Poisson's ratio, k is
the thermal conductivity, α is the coefficient of thermal expansion and E is Young's modulus. Materials with higher values of
the parameters R and R' will have higher thermal stress resistance
than those with lower values of these parameters. A review of
some thirty other thermal stress resistance parameters was presented some time ago. The tensile strength, S_t, is present in R and R'
because of the high ratio of compressive to tensile strength of
brittle materials, which makes them most susceptible to failure in
tension.

Radiation represents an important mechanism of heat transfer,
especially at high temperatures. The incidence of thermal stress
failure under conditions of radiative heat transfer is receiving
increased attention lately as the direct result of the importance
of solar radiation as an energy source. An analysis of the thermal stress failure of brittle materials opaque to incident black-
body radiation was presented some time ago [3]. This analysis was
extended to include materials opaque above a given wavelength of

incident black-body radiation and completely transparent below
this wavelength [4].

The absorption properties of many materials are such that incident radiation transmitted through the surface is absorbed throughout the thickness. This results in the so-called "thermal trapping"
effect, in which the temperature in the interior of a material can
greatly exceed the temperature of the ambient environment. An
analysis of the thermal stresses which result from this type of
heat transfer was performed recently [5] for a flat plat symmetrically heated by normally incident radiation and cooled by convection with heat transfer coefficients h = 0 and ∞.

The purpose of this paper is to report these latter results,
together with results for the thermal stresses in a plate asymmetrically heated by radiation and cooled by convection again for h
values of 0 and ∞. In this paper, special attention is paid to
the interaction between the coefficient of thermal expansion and
the other material properties which control the magnitude of thermal stress. Only the principal analytical results will be high-
lighted. For the mathematical details the reader is referred to
the original report [5,6].

ANALYSIS

The analysis of thermal stresses first requires obtaining the
solutions for the temperature distributions. From these, expressions are derived for the thermal stresses. By equating the maximum tensile thermal stress to the tensile strength, an expression
can be obtained for the maximum incident radiation heat flux to
which the material can be subjected without causing failure.

For the anslysis, the following simplifying assumptions were
made. First, reflectivity of the material of the plate is sufficiently low that the effect of multiple internal reflections can
be neglected. secondly, the plate will be assumed to be at low
enough temperature that the maximum value of thermal stress is
reached before the plate becomes sufficiently hot that re-emission
of the absorbed radiation must be taken into account. Indeed, the
validity of this latter assumption is easily demonstrated with
numerical examples [5]. Finally, it was also assumed that the
emissivity and absorptivity of the plate are independent of wavelength. If required, the spectral dependence of these properties
can easily be taken into account in the analysis.

For the various types of radiative heating and convective
cooling conditions and the rates of internal heat generation,
the general approach used for the solutions and the expressions
for the maximum value of tensile thermal stresses are as follows.

Symmetric Radiative Heating and Convective Cooling

For normally incident symmetric thermal radiation, the rate of internal heat generation (g''') within the plate can be derived to be:

$$g''' = 2\mu\varepsilon q_o e^{-\mu a} \cosh(\mu x) \tag{1}$$

where μ is the absorption coefficient, ε is the emissivity ($\varepsilon = 1 - r$, where r is the reflectivity), q_o is the intensity of the incident heat flux at each side of the plate, a is the half-thickness of the plate and x is the through-the-thickness coordinate with $x = 0$ at the center of the plate.

Derivation of the transient temperature (T) requires solution of the differential equation [7]

$$\partial^2 T/\partial x^2 + g'''(x)/k = (1/\beta)\partial T/\partial t \tag{2}$$

where k is the thermal conductivity, t is the time, and β is the thermal diffusivity.

The thermal stresses in the plane of the plate are obtained by substitution of the solution for the temperature in [8]:

$$\sigma_{y,z} = \frac{\alpha E}{1 - \nu} [-T + \frac{1}{2a} \int_{-a}^{a} T dx + \frac{3x}{2a^3} \int_{-a}^{a} Tx dx] \tag{3}$$

For the two cooling conditions, the initial and boundary conditions used for the solution of Eq. (2) and the maximum tensile stresses and maximum permissible heat flux are as follows:

Heat transfer coefficient h = 0. Initial and boundary conditions:

$$T(x,0) = T_o; \quad \partial T/\partial x(0,t) = \partial T/\partial x(a,t) = 0 \tag{4}$$

For these conditions the maximum tensile thermal stresses which occur at $x = 0$ and $t = \infty$ can be derived to be:

$$\sigma_{y,z} = \frac{\alpha E \varepsilon q_o e^{-\mu a}}{(1 - \nu)k} \{\frac{2}{\mu} + (\frac{a}{3} - \frac{2}{a\mu^2}) \sinh(\mu a)\} \tag{5}$$

In general, by equating the maximum tensile thermal stress to the tensile strength, Eq. (5) can be rearranged into an expression for the maximum permissible heat flux, q_{max}, such that thermal stress failure is avoided. For the limiting cases of an optically thick ($\mu a \to \infty$) and optically thin ($\mu a \ll 1$) plate, simple analytical expressions for q_{max} can be obtained:

$$q_{max} = 6S_t(1 - \nu)k/\alpha E \varepsilon a \qquad (\mu a \to \infty) \qquad (6a)$$

$$q_{max} = 180S_t(1 - \nu)k/7\alpha E \varepsilon \mu^3 a^4 \qquad (\mu a \ll 1) \qquad (6b)$$

Heat transfer coefficient h = ∞. For this case the initial and boundary conditions are:

$$T(x,0) = T_o; \quad \partial T/\partial x\,(0,t) = 0 \qquad\qquad (7a)$$

$$T(a,t) = T(-a,t) = T_o \qquad\qquad (7b)$$

The tensile thermal stresses are a maximum at t = ∞ and x = -a,a and can be derived to be:

$$\sigma_{y,z}(a,\infty) = \frac{4\alpha E(\mu a)\varepsilon q_o e^{-\mu a}}{(1 - \nu)ka^2} \cosh(\mu a) \sum_{n=o}^{\infty} \{(\mu^2+\lambda_n^2)(a\lambda_n^2)\}^{-1} \qquad (8)$$

where $\lambda_n = (n + 1/2)\pi/a$.

In analogy to Eq. (6), the maximum permissible radiation heat flux q_{max} for the optically thick and optically thin plate can be derived to be:

$$q_{max} = \infty \qquad (\mu a \to \infty) \qquad (9a)$$

$$q_{max} = \frac{S_t(1-\nu)k\pi^4}{64\alpha E\mu\varepsilon a^2} \qquad (\mu a \ll 1) \qquad (9b)$$

Asymmetric Radiation Heating and Convective Cooling

The thermal stresses will be given specifically for a plate subjected to incident radiation on one face (x = -a) and convective cooling on the opposite face (x = a).

For this case the rate of internal heat generation is:

$$g''' = q_o \varepsilon \mu e^{-\mu(x + a)} \qquad\qquad (10)$$

Heat transfer coefficient h = 0. The initial and boundary conditions are

$$T(x,0) = T_o; \quad \partial T/\partial x(-a,t) = \partial T/\partial x(a,t) = 0 \qquad (11)$$

The solutions for the stresses indicate that the tensile thermal stresses reach their maximum value at $t = \infty$, but that the position of the maximum tensile stress can occur anywhere within the plate, depending on the value of optical thickness. The complete solution for the thermal stresses at $t = \infty$ as a function of the coordinate x is:

$$
\sigma_{y,z} = \frac{\alpha E q_o \epsilon}{(1-\nu)k} \left[\frac{e^{-\mu(x+a)}}{\mu} - \frac{(x+a)^2}{4a}(1-e^{-2\mu a}) + (x+a) \right]
$$

$$
+ \frac{\alpha E q_o \epsilon}{(1-\nu)k} \left[\frac{e^{-2\mu a} - 1}{2\mu^2 a} + \frac{a}{3}(1 - e^{-2\mu a}) - a \right]
$$

$$
+ \frac{3\alpha E q_o \epsilon x}{2(1-\nu)a^3 k} \left\{ \frac{1}{\mu^3} \left[e^{-2\mu a}(\mu a + 1) + (\mu a - 1) \right] \right.
$$

$$
\left. + \frac{a^3}{3}(1 - e^{-2\mu a}) - \frac{2a^3}{3} \right\} \tag{12}
$$

For the optically very thick and very thin plates the maximum tensile stresses occur at $x = 0$ and at $x = -a$ respectively and are given by:

$$
\sigma_{y,z}(0,\infty) = \frac{\alpha E q_o \epsilon a}{12(1-\nu)k} \qquad (\mu a \to \infty) \tag{13a}
$$

$$
\sigma_{y,z}(-a,\infty) = \frac{\alpha E q_o \epsilon \mu a^2}{(1-\nu)k} \qquad (\mu a \ll 1) \tag{13b}
$$

The expressions for the maximum permissible heat flux become

$$
q_{max} = 12 S_t (1-\nu)k/\alpha E \epsilon a \qquad (\mu a \to \infty) \tag{14a}
$$

$$
q_{max} = S_t (1-\nu)k/\alpha E \epsilon \mu a^2 \qquad (\mu a \ll 1) \tag{14b}
$$

Heat transfer coefficient h = ∞. The initial and boundary conditions are:

$$
T(x,0) = T(a,t) = T_o; \quad \partial T/\partial x(-a,t) = 0
$$

The tensile thermal stresses have their maximum value at $x = -a$ and $t = \infty$, and are given by

$$\sigma_{y,z} = -\frac{\alpha E}{1-\nu} \sum_{n=odd}^{\infty} \frac{16q_o\varepsilon\mu a}{n^2\pi^2 k} \frac{\mu + (n\pi/4a)e^{-2\mu a}}{\mu^2 + (n\pi/4a)^2}$$

$$+ \frac{\alpha E}{1-\nu} \sum_{n=odd}^{\infty} \frac{8q_o\varepsilon\mu}{n^2\pi^2 k} \frac{\mu + (n\pi/4a)e^{-2\mu a}}{\mu^2 + (n\pi/4a)^2} \frac{1}{\lambda_n}(-1)^{\frac{n-1}{2}}$$

$$- \frac{3\alpha E}{a(1-\nu)} \sum_{n=odd}^{\infty} \frac{8q_o\varepsilon\mu}{n^2\pi^2 k} \frac{\mu + (n\pi/4a)e^{-2\mu a}}{\mu^2 + (n\pi/4a)^2} \left[\frac{a}{\lambda_n}(-1)^{\frac{n-1}{2}} - \frac{1}{\lambda_n^2}\right] \quad (15)$$

where $\lambda_n = n\pi/4a$ with $n = 1, 3, 5, 7 \ldots$

For the optically thick and optically thin plates Eq. (15) can be simplified to:

$$\sigma_{y,z}(-a,\infty) = 0 \qquad\qquad (\mu a \to \infty) \qquad\qquad (16a)$$

$$\sigma_{y,z}(-a,\infty) = \frac{0.2719\alpha Eq_o\varepsilon\mu a^2}{(1-\nu)k} \qquad (\mu a \ll 1) \qquad\qquad (16b)$$

which yield for the maximum incident heat flux:

$$q_{max} \to \infty \qquad\qquad (\mu a \to \infty) \qquad\qquad (17a)$$

$$q_{max} = \frac{3.677S_t(1-\nu)k}{\alpha E\varepsilon\mu a^2} \qquad (\mu a \ll 1) \qquad\qquad (17b)$$

NUMERICAL RESULTS AND DISCUSSION

For the symmetrically heated place, Fig. 1 shows the maximum tensile thermal stress as a function of the optical thickness, μa. For $h = \infty$, the stresses are identically equal to zero for $\mu a = 0$ and ∞, with a maximum at $\mu a = 1.34$. For $\mu a = 0$, no thermal stress will arise, because no heat is absorbed. For $\mu a = \infty$, the radiation is absorbed in the immediate surface and removed by the convection medium. Again, no heating of the plate occurs. At $\mu a = 1.34$, the thermal trapping effect is a maximum with the corresponding highest values of temperature increase and thermal stresses within the plate. These results suggest that in general for a highly efficient solar collector ($\mu a \to \infty$), the heat should be removed from the plate at the same side as the incident radia-

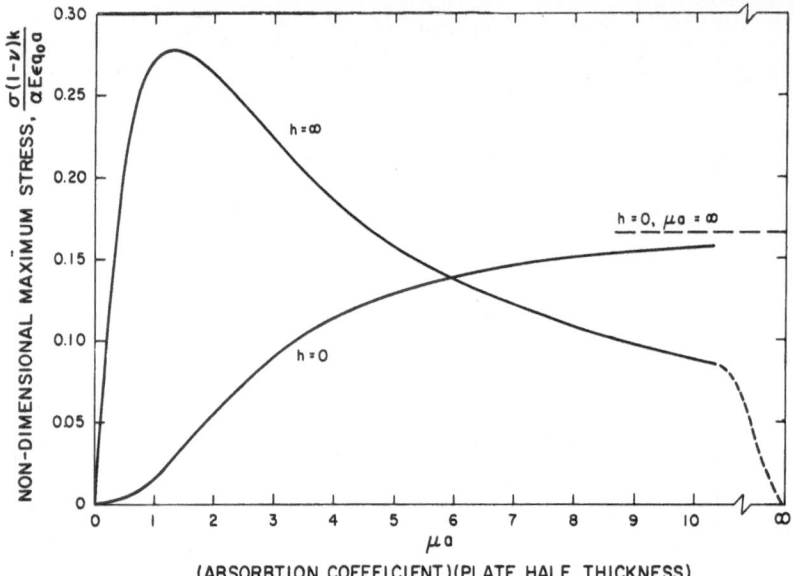

Fig. 1. Magnitude of maximum tensile thermal stresses as a func-
         tion of optical thickness in flat plate symmetrically
         heated by thermal radiation and cooled by convection.

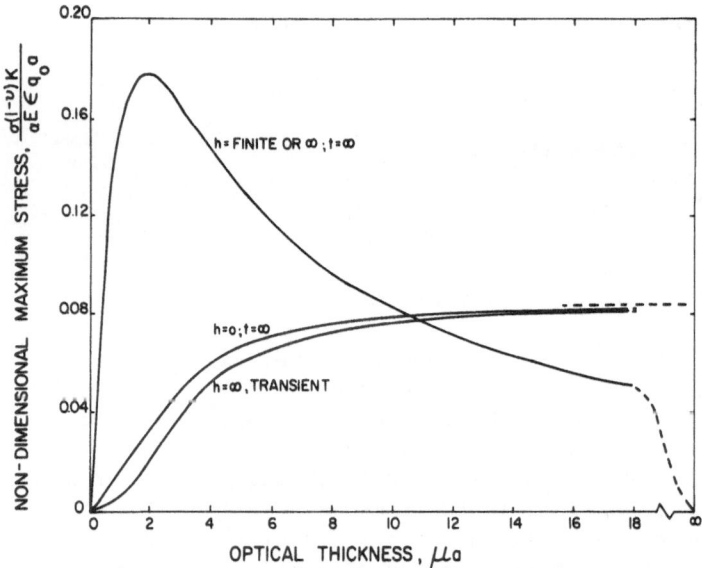

Fig. 2. Magnitude of maximum tensile thermal stresses as a func-
         tion of optical thickness in flat plate asymmetrically
         heated by thermal radiation on one side and cooled by con-
         vection on the opposite side.

tion. For h = 0, the stresses are zero for μa = 0 and increase
monotonically with increasing value of μa.

Figure 2 shows the maximum values of tensile stress as a function of the optical thickness of the plate heated on one side by
the incident radiation and cooled on the other side by convection.
Qualitatively, Fig. 2 looks similar to Fig. 1. The magnitude of
the stresses for the asymmetrically heated plate, however, is much
less than that for the symmetrically heated plate. The principal
reason for this is that in the asymmetrically heated flat plate,
the non-uniform thermal expansions in part can be accommodated by
bending.

For h = ∞ and μa = 0, no stresses occur because no heat is
absorbed. For μa = ∞ and h = ∞, although the heat is transmitted
through the plate, no stresses occur because at t = ∞ the temperature distribution is linear, which results in zero thermal stress.
For h = 0, the stress is zero for μa = 0 and increases monotonically with increasing μa. Included in Fig. 2 are the values of the
maximum transient stresses for h = ∞. For these, the reader is
referred to the original study [6] for further details.

Attention can now be focused on the role of the coefficient
of thermal expansion and its interaction with the other material
properties establishing the magnitude of the thermal stress resistance of semi-absorbing ceramics. It may be noted that the various expressions for q_{max} are a function of a numerical constant,
the thickness of the plate, and a number of material properties.
For a given heating and cooling condition and plate thickness,
thermal stress resistance is governed only by the material properties. For the purpose of selection of the material with optimum
thermal stress resistance, so-called "thermal-stress-resistance"
parameters or figures-of-merit can be separated from the expressions for q_{max} as follows:

$$\frac{S_t(1-\nu)k}{\alpha E \epsilon \mu} ; \quad \frac{S_t(1-\nu)k}{\alpha E \epsilon \mu^3} ; \quad \frac{S_t(1-\nu)k}{\alpha E \epsilon} \tag{18}$$

Materials with higher values of these figures-of-merit have better
thermal stress resistance than materials with lower values. For
this reason, high thermal stress resistance is associated with
high values of tensile strength and thermal conductivity, coupled
with low values of the coefficient of thermal expansion, Young's
modulus, emissivity and absorption coefficient. For optically
thin materials (μa << 1), thermal stress resistance is inversely
proportional to μ or μ^3 depending on the heating and cooling condition, but is independent of μ for optically thick materials and
h=0. Of course, since the objective of solar collectors is to ab-
sorb as much radiation as possible, the values of ε and μ should

be non-zero. For this reason, trade-offs in design and material properties may be required.

In terms of the objective of the present conference, it is of interest to note that in establishing the values for the figures-of-merit in Eq. (18), the coefficient of thermal expansion interacts with as many as six other material properties, including optical, thermal as well as mechanical properties. As far as the present writers are aware, no figure-of-merit for materials in such engineering fields as aerospace, thermoelectricity, etc. contains as many dissimilar material properties such as those given by Eq. (18). Clearly, the coefficient of thermal expansion will play a significant role in determining the feasibility of high-intensity solar energy collectors.

ACKNOWLEDGMENTS

This study was conducted as part of a research program on the thermal properties of engineering ceramics for structural applications supported by the Office of Naval Research under contract N00014-78-C-0413.

REFERENCES

1. W. D. Kingery, J. Amer. Ceram. Soc., 38 (1955) 3.
2. D. P. H. Hasselman, Ceramurgia, 4 (1978) 147.
3. D. P. H. Hasselman, J. Amer. Ceram. Soc., 46 (1963) 229.
4. D. P. H. Hasselman, J. Amer. Ceram. Soc., 49 (1966) 103.
5. D. P. H. Hasselman, J. R. Thomas, Jr., M. P. Kamat and
        K. Satyamurthy, J. Amer. Ceram. Soc., 63 (1980) 21.
6. J. R. Thomas, Jr., J. P. Singh and D. P. H. Hasselman,
        J. Amer. Ceram. Soc. (in press).
7. H. S. Carslaw and J. C. Jaeger, Conduction of Heat in Solids,
        Oxford at the Clarendon Press (1960).
8. B. Boley and J. Wiener, Theory of Thermal Stresses, John Wiley
        and Sons, New York (1960).

AN ANALYSIS OF THERMAL EXPANSION

AND MELTING IN ALKALI HALIDES

L. L. Boyer

Naval Research Laboratory

Washington D.C. 20375

An analysis of thermal expansion and melting is presented which predicts, to within a reasonable accuracy expected from the approximations of the theory, the melting temperature (T_M) and the thermal expansion up to and including the discontinuity at T_M. Based on first-principles equation of state calculations for alkali halides, the lattice constant is shown to have a $(T_c - T)^{1/2}$ behavior in the high temperature limit. This implies that the thermal expansion coefficient diverges as $T \to T_c$ and hence the isothermal bulk modulus goes to zero at T_c. A least squares analysis of published thermal expansion data for NaCl using this analytic form suggests that $T_M \lesssim T_c \gtrsim 1.1 \ T_M$, but more accurate data is needed to fix T_c within this range.

INTRODUCTION

Recent work on first-principles equation of state calculations for alkali halides[1,2] have shown that a reasonably accurate equation of state can be obtained without using experimentally determined parameters. One of the more interesting aspects of this work is the prediction of a kind of thermodynamic instability which seems to play a role in explaining the phenomenon of melting. In this paper we compare the predictions of the theory, which pertain to thermal expansion and melting, with experimental results for NaCl.

In spite of a large and continuing effort to understand how and why solids melt, a generally accepted theory of melting is yet to emerge. Most theories, however, have a common feature in that they rely upon some kind of instability of the solid phase as a driving mechanism for the melting transition.[3] To illustrate the kinds of instabilities which have been proposed I have selected the following: Lindeman[4] (critical mean square displacement); Herzfeld and Goeppert-Mayer[5] and Kane[6] (vanishing of β_T, the isothermal bulk modulus); Born[7] (vanishing of a shear elastic modulus); Gorecki[8] (vacancy formation); Kuhlmann-Wilsdorf[9] (generation of glide dislocations); and Cotterill[10] (surface related generation of Shockley partial dislocations). As pointed out in

Ref. 1, the present results tend to support the ideas of Herzfeld and Goeppert-Mayer. However, in this paper we emphasize that the different types of instabilities can take on a cause and effect relationship; to say that one type of instability is the cause of melting is not telling the whole story.

In order to assess the role of the $\beta_T \to 0$ instability in melting we compare calculated and measured results for thermal expansion in NaCl. Remember that β_T is related to the coefficient of thermal linear expansion, α, by the thermodynamic relationship[11]

$$\frac{1}{\beta_T} = \frac{1}{\beta_s} + \frac{TV9\alpha^2}{C_p} \tag{1}$$

where β_s is the adiabatic bulk modulus, and C_p is the heat capacity at constant pressure. Thus, the isothermal compressibility, $1/\beta_T$, diverges $(\beta_T \to 0)$ when α, or strictly speaking, α^2/C_p, diverges. I have selected NaCl for this study because of the relatively large amount of high temperature thermal expansion data available. To facilitate subsequent discussion let T_M be the melting temperature and T_c be the temperature for which $\beta_T = 0$. It is natural to expect α to diverge as $T \to T_M$ because α for the solid-liquid system at T_M is necessarily infinite due to the volume discontinuity that accompanies melting. However, the question is, what is the limiting behavior of α for the solid as $T \to T_M$, and to what extent can it be described by an instability of the perfect crystal in which $\beta_T \to 0$ as $T \to T_c$. We shall see that the experimental results in a large measure support the $\beta_T \to 0$ picture of melting but the data are not sufficiently accurate to draw firm conclusions. Specifically, a greater-than-linear increase in α, which can be interpreted as the onset of the $\beta_T \to 0$ instability, is apparent in most high temperature thermal expansion measurements, but the magnitude of $T_c - T_M$ is still open to question.

Theory

The equation of state for a crystal whose structure is given by its volume, V, may be written[12]

$$P + du/dV = f(V,T) \tag{2}$$

where

$$f(V,T) \equiv V^{-1} \sum_i \gamma_i \{ h\nu_i/2 + h\nu_i/[\exp(h\nu_i/KT) - 1] \}, \tag{3}$$

$$\gamma_i \equiv (-V/\nu_i)\,(d\nu_i/dV), \tag{4}$$

P is the external pressure, u is the electronic ground state energy of the static lattice, $-f(V,T)$ is a pressure due to the vibration of the ions, and the quantities, $\nu_i = \nu_i(V)$, are the classical frequencies of small amplitude oscillations about the lattice sites given by V. The derivation of Eq. (2) makes use of the so called quasi-harmonic approximation which assumes that the vibrational energy levels of the system are those of independent quantum harmonic oscillators with frequencies $\nu_i(V)$. In this work the electron pressure (du/dV) and the phonon pressure $(f(V,T))$ were calculated from pair potentials which were themselves calculated using the theory of Gordon and Kim.[13] Details of the method of the calculation are the same as that described in Ref. 2, except that for this work the exponential form $Ce^{(\beta r + \delta r^2)}$, rather than $Ce^{\beta r}$, was used in the numerical fit to the short range parts of the calculated pair potentials.

Results

The calculated equation of state for NaCl is presented graphically in Fig. 1, where the electron and phonon pressures are plotted as a function of the lattice constant, a. The electron pressure (du/dV) is independent of T and is shown with a discontinuity where the scale changes by a factor of 10. The phonon pressure $(f(V,T))$ is plotted for several selected temperatures from 0 K to 1400 K. For zero external pressure the equilibrium lattice constants, for the selected temperatures, are given by the points on the abscissa where the electron pressure curve intersects the phonon pressure curves. Notice that some of the phonon pressure curves intersect the electron pressure curves in two places (the first (second) intersection corresponds to a minimum (maximum) in the free energy) while above a certain critical temperature, T_c, there is no such intersection and the lattice is therefore unstable. Exactly at T_c the two intersections coalesce and the slopes of the electron and phonon pressure curves are equal at this point. It follows then that $\beta_T = 0$ at T_c.

There are three points I wish to make about these results for NaCl and the alkali halides in general that bear on melting. First of all, the calculated values of T_c for the alkali halides are in reasonable accord with the melting temperatures. For NaCl the agreement is nearly perfect ($T_c = 1070$ K, $T_m = 1073$ K), but such good agreement is fortuitous. Generally speaking the agreement for other materials is $\sim$ 20% with progressively worse agreement for those compounds that have the largest difference in the sizes of the constituent ions (up to $\sim$ 50% for the Li compounds); this trend has been attributed to a worsening of the pair potential approximation.[2] Furthermore, those compounds which have a good agreement between T_c and T_M have a similarly good agreement for other equation of state properties, such as thermal expansion. (This trend is further discussed and refined to include, qualitatively, the effect of anharmonic corrections in Ref. 2.) The fact that the calculations were accomplished without experimental input,[14] adds credence to the correlation between T_c and T_M.

The second point involves the three small arrows on the abscissa of Fig. 1. They point to the experimentally determined values of the lattice constant of the solid at 0 K and T_M, and an effective value based on the volume of the liquid[15] at T_M. As we shall see (Fig. 2) there is an appreciable amount of uncertainty in the value for the solid at T_M. From Fig. 1 we see that the calculated values of a for the solid, at 0 K and T_c, are in good agreement with the corresponding measured values. In order to compare with the value for the liquid we must first note the reason for the sharp increase in the phonon pressure curves at $\sim$ 6.2 Å. This is caused by the decrease in the shear elastic constant $C_{11} - C_{12}$, with increasing volume, which according to the theory reaches zero at 6.25 Å. Thus, from Fig. 1, we see that it is this decrease in $C_{11} - C_{12}$ that increases the corresponding γ_i in Eq. (3) and <u>causes</u> the $\beta_T \rightarrow 0$ instability to occur before the lattice can be thermally expanded to the critical volume, $(6.25 \text{ Å})^3$, where $C_{11} - C_{12} = 0$. We shall return to this point later. It is natural to equate this volume

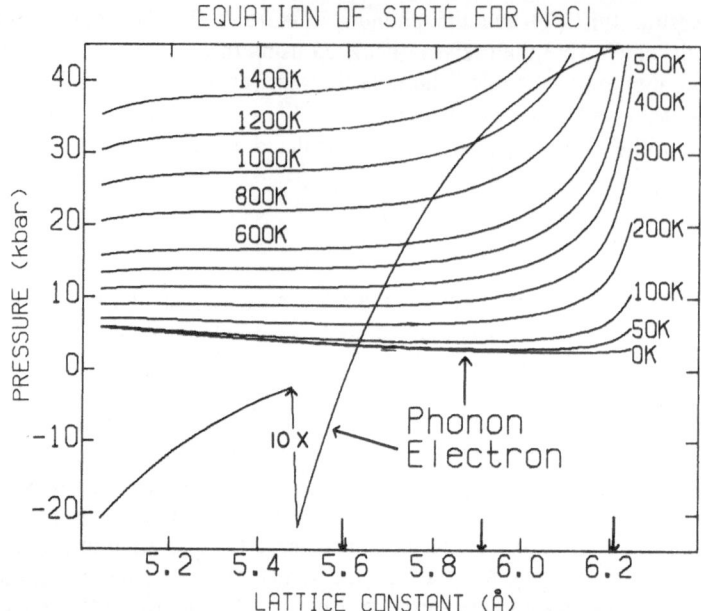

Fig 1. Calculated electron and phonon pressures as a function of lattice constant for NaCl. On the abscissa three arrows mark the experimental lattice constant at 0 K, the melting temperature, and an effective lattice constant for the liquid at the melting temperature.

to that of the liquid at T_M because at this volume the static lattice, like the liquid, would have no resistance to elastic shear. From Fig. 1 we see that this volume is also in good agreement with the measured volume of the liquid at T_M.

Finally, we turn our attention to the thermal expansion of NaCl between room temperature and T_M. In Fig. 2 the calculated % thermal expansion is compared with measured results from several sources[16-19] as tabulated in Ref. 20. The calculated results fall essentially within the experimental error expected from the scatter in the data. As $T \to T_c$ the theory gives a limiting value of 5.4 %. As noted earlier, the corresponding measured limiting value is clearly quite uncertain. Beyond about 950 K the scatter in the data increases markedly; at about 1050 K, where the data stop, the results range from $\sim$ 3.9 to 4.9. This uncertainty in the high temperature data is even more pronounced in the fig of Ref. 20, where more data are plotted. Obviously, not very much can be said about the analytic behavior of α in the limit as $T \to T_M$ from a visual inspection of the data in Fig. 2. A misleading picture can be obtained from the common practice of reporting experimental results in terms of a polynomial fit to the

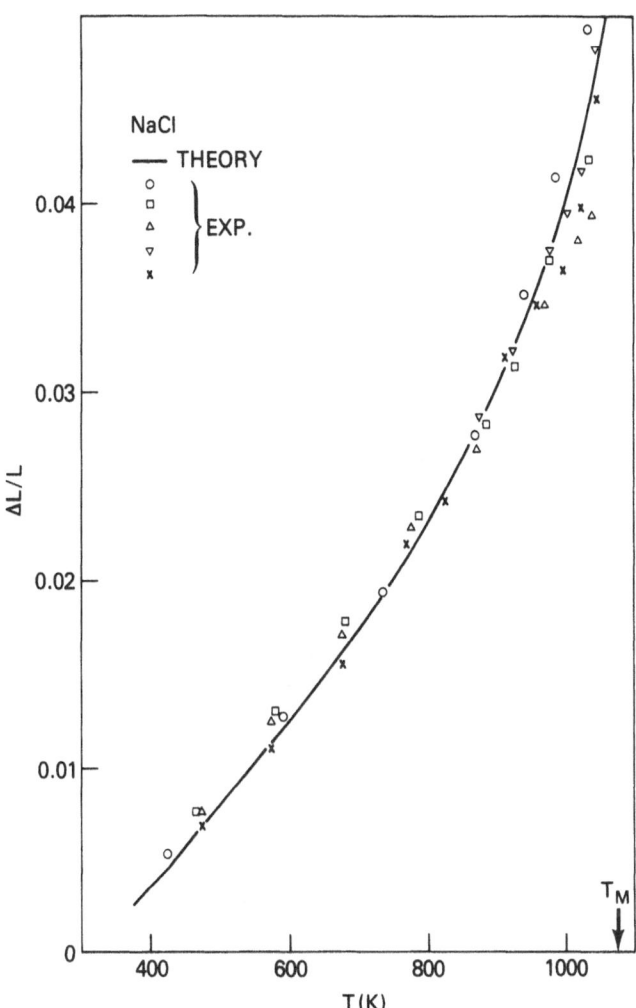

Fig. 2. Comparison of the calculated thermal expansion at high temperatures with selected experimental data (0 - Ref. 16, □ - Ref. 17, Δ - Ref. 18, ▽ - Ref. 19, and × - Ref. 19) as tabulated in Ref. 20.

data. In particular, Srivastava and Merchant[21] report the coefficients of only a quadratic fit to the lattice constant which they claim is valid from room temperature to near melting. This, of course, restricts α to be nearly linear over the entire temperature range. At the same time they neglect to mention the substantial greater-than-linear increase in α seen by earlier workers.[17-19]

Notwithstanding the uncertainties in the high temperature data, we will attempt to analyze some of the experimental results in terms of the present theory; but first, we need to determine the theoretical analytic behavior of a and hence α as $T \to T_c$. For this we make use of the fact that at high temperatures the phonon pressure, for fixed a, is essentially proportional to T. This can be seen from Fig. 1 or by expanding the exponential factor in Eq. (3). Also, in the critical region, the a dependence of the electron and phonon pressure curves can each be approximated with a quadratic expression in a; the only difference is that the zeroth order term in the phonon pressure is proportional to T. The equilibrium lattice constant is a solution of the resultant quadratic equation. Therefore, at high temperatures (near T_c) $a(T)$ may be written

$$a(T) = a_0 + a_1(T_c - T)^{1/2}. \tag{5}$$

I have used this expression to analyze the data of Smirnov et. al. and Fishmeister.[22] The result of the fit to the Smirnov data (above 550 K) is shown in Fig. 3 along with the data and a quadratic fit. Higher order polynomial fits to the data were found to merely produce more oscillations about the fit obtained from Eq. (5). The Fishmeister data was more amenable to polynomial fitting than the Smirnov data, but again, the polynomial fits were found to oscillate about that obtained from Eq. (5). The optimum value of T_c obtained from the Smirnov data was precisely the melting temperature (1073 K). While such precise agreement may be partly accidental, the optimal value for the Fishmeister data, which shows a much smaller increase in a at high temperatures, was found to be $\sim$1180 K, only 10% higher than the melting temperature.

It should be noted that this type of high temperature thermal expansion behavior is seen in x-ray measurements as well as those which measure macroscopic volume. In particular, Smirnov et al. find good agreement between results determined from x-ray diffraction and pycnometer measurements.

Conclusions

The above results show that the $\beta_T \to 0$ instability, predicted by first-principles calculations of the equation of state of alkali halides, provides an explanation of the phenomenon of melting which is not in any substantial disagreement with experimental results. The results strongly suggest that this type of thermodynamic instability should at least play a role in any more complete theory of melting.

In attempting to assess the significance of this role one should be careful to distinguish between the cause of melting and the explanation of the properties of the solid at the melting point. One should not expect a theory which does not account for lattice imperfections (vacancies, dislocations, surfaces, etc.) to give a complete description of the properties of the solid all the way to the melting point. If such were the case we

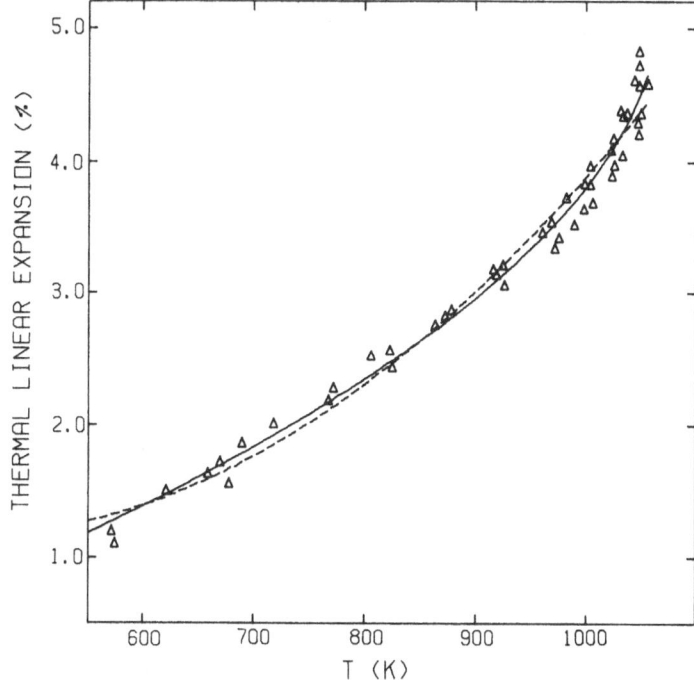

Fig. 3. Least Square fits to Data from Ref. 19 (as tabulated in Ref. 20) using Eq. (5) (solid line) and a 2nd order polynomial in T (dashed line).

would have $T_c = T_M$. We have seen that the experimental results for thermal expansion are not accurate enough to establish a firm value for T_c, but the available data indicate that $T_c - T_M \gtrsim 100\text{K}$.

There does not appear to be any one type of instability that can be singled out as the cause of melting. In fact the various types of instabilities previously proposed as the cause of melting are interrelated. We have seen how the $\beta_T \to 0$ instability is itself caused by the decrease in $C_{11} - C_{12}$. It is reasonable to further expect that the softening of the lattice, due to the diminishing value of β_T, may promote another type of instability (presumably in the form of some sort of lattice imperfection) to the critical stage before β_T reaches zero.

On the other hand, the concept of equilibrium as a balance between a pressure due to the electronic energy and one due to the vibrational energy is one which should be just as valid for a solid with lattice imperfections. It is not difficult to imagine that a similar calculation for a finite crystal, or one with free surfaces, would produce a slightly lower value of T_c.

Acknowledgments

This work has benefited from numerous discussions with J.R. Hardy and from stimulating correspondence with J.L. Tallon.

References

1. L. L. Boyer, Phys. Rev. Lett. 42:584 (1979); ibid. 45:1858 (1980).
2. L. L. Boyer, Bull. Am. Phys. Soc. 24:458 (1979), and Phys. Rev. B, to be published.
3. An exception can be found in the work of Tallon et al. (J. L. Tallon, W. H. Robinson, and S. I. Smedley, Phil. Mag 36:741 (1977) and Nature 266:337 (1977)) who propose a two phase theory in which the liquid is characterized as a "virtual solid" with no resistance to elastic shear.
4. F. A. Lindemann, Z. Phys. 11:609 (1910).
5. K. F. Herzfeld and M. Goeppert-Mayer, Phys. Rev. 46:995 (1934).
6. G. Kane, J. Chem. Phys. 7:603 (1939)
7. M. Born, J. Chem. Phys. 7:591 (1939).
8. T. Gorecki, Z. Metallk 65:426 (1974).
9. D. Kuhlmann-Wilsdorf, Phys. Rev. 140:A1599 (1965).
10. R. M. J. Cotterill, Phil Mag. 32:1283 (1975).
11. K. Huang, *Statistal Mechanics* (Wiley, New York-London, 1963), Eq. 1.24.
12. M. Born and K. Huang, *Dynamical Theory of Crystal Lattices* (Oxford Univ. Press, London, 1954).
13. R. G. Gordon and Y. S. Kim, J. Chem. Phys. 56:3122 (1972).
14. The only experimental quantities which enter the calculation are Planck's constant, Boltzmann's constant, the mass of the electron, the charge of the electron, the mass of the nuclei, and the atomic number.
15. G. J. Janz, F. W. Dampier, G. R. Lakshminarayanan, P. K. Lorenz, and R. P. T. Tomkins, NSRDS-National Bureau of Standards, 15 (1968).
16. E. Laredo, J. Phys. Chem. Sol. 9:401 (1963).
17. H. F. Fishmeister, Acta Cryst. 9:416 (1956).
18. F. D. Enck and J. L. Dommel, J. Appl. Phys. 36:839 (1965).
19. M. V. Smirnov, M. M. Vasipevskaya, and G. V. Vupov, Trudy Instituta Elektro-khimic, Akademiya Nauk SSSR, Uralśkii Filial, 13:82 (1970).
20. "Thermal Expansion of Nonmetallic Solids," Vol. 13 of Thermophysical Properties of Matter, ed. by Y. S. Touloukian and C. Y. Ho (Plenum, New York-Washington, 1977).
21. K. K. Srivastava and H. D. Merchant, J. Phys. Chem. Solids 34:2069 (1973).
22. Again, I have used the data as it has been tabulated in Ref. 20. The Fishmeister data was obtained by integrating plotted data for α. The Smirnov data shown in this paper, identified as curves 41, 42, 44, 45, 46, and 48 in Ref. 20, was taken directly from plotted results in Ref. 19.

CALCULATION OF THERMAL EXPANSION IN INSULATING

AND CERAMIC MATERIALS

A. R. Ruffa

Naval Research Laboratory

Washington, D.C. 20375

ABSTRACT

A method of calculating thermal expansion in insulating
and ceramic materials is presented which makes use of a theory
of thermal expansion in terms of the quantum mechanical solutions
of the Morse potential. In this model, the localized interatomic
potential solutions, obtained from the appropriate Morse potential, are combined with the Debye model to give a localized-
continuum description of thermal expansion. A set of empirical
rules is developed for characterizing the interatomic potential in
terms of the parameters of the Morse potential. These are then
applied to the quantitative calculation of thermal expansion in
the alkali halide crystals and a group of binary high temperature
materials with the aid of the known crystal structures, compressi-
bilities, and Debye temperatures of these materials. Good agreement between theoretical and experimental values is obtained for
these materials at temperatures ranging from 0°K to values near
their melting points. Using further empirical treatment, thermal
expansion in a larger group of binary and complex ternary materials
is calculated using no experimental input other than the chemical
formula. The agreement with experiment is again generally good,
though less accurate than when experimental input is used. The
results indicate that this approach is capable of predicting the
thermal expansion of a wide range of materials with little or no
experimental input and no adjustable parameters. The limitations
of this method for certain special cases is also discussed.

INTRODUCTION

This presentation will report on recent progress which has been made in predicting the thermal expansion in insulating and ceramic materials with little or no input of experimental data and no adjustable parameters. The calculations make use of a simple localized-continuum model of thermal expansion which was originally applied to simple metals.[1] Basically, the model accounts for the anharmonic properties by means of a model interatomic potential, the Morse potential. The exact quantum mechanical solutions for this potential are known, and an exact thermal average of the displacement from equilibrium can then be performed in closed form. This in turn can then be averaged over the Debye frequency spectrum, which assumes a continuum of normal mode frequencies with a quadratic frequency dependence. This approach was found to work well for the simple metals.

The Morse potential used in these calculations has the simple form:

$$V(r) = D(1 - e^{-a(r-r_n)})2$$

dependent upon only two parameters, the depth D and the inverse width a. The equilibrium distance in insulators is set equal to the nearest-neighbor distance, r_n. The thermal expansion formulas which one obtains from this approach are summarized below.

$$\Delta \ell / \ell = (3kT/2ar_n D)(T/\theta_D)^3 f(x_D)$$

$$f(x_D) = \int_0^{x_D} \frac{x^3 dx}{e^x - 1}$$

$$\alpha(T) = (3k/2ar_n D)(T/\theta_D)^3 g(x_D)$$

$$g(x_D) = \int_0^{x_D} \frac{x^4 e^x dx}{(e^x - 1)^2}$$

$$(\Delta \ell / \ell)_1 = (3kT/2ar_n D)(kT/4D)(T/\theta_D)^3 f_1(x_D)$$

$$f_1(x_D) = \int_0^{x_D} \frac{x^4(1 + e^x)dx}{(e^x - 1)^2}$$

$$\alpha(T)_1 = (3k/2ar_n D)(kT/2D)(T/\theta_D)^3 g_1(x_D)$$

$$g_1(x_D) = \int_0^{x_D} \frac{x^5 e^x(1 + e^x)dx}{(e^x - 1)^3}$$

These formulas arise from expanding the contributions to the thermal expansion in a power series in D^{-1}. The first two lines give the leading contributions to the percent thermal expansion and the coefficient of thermal expansion, respectively. The latter has the functional form of the Debye specific heat and the results are scaled on a temperature scale by the value of the Debye temperature $\theta_D(X_D = \theta_D/T)$. The next two lines give the second order contributions to both quantities. At high temperatures, the second order contribution to the coefficient of thermal expansion is a fractional linear correction to the leading term proportional to kT/D. Since the leading term is inversely proportional to ar_nD, there is a simple connection in this model between the shape of the coefficient of thermal expansion curve and the parameters of the Morse potential. The contributions shown here are usually enough to characterize the thermal expansion up to temperatures approaching the melting temperature.

In order to apply these formulas to real materials, it is necessary to make a connection between the parameters of the Morse potential and the characteristics of the actual interatomic potential. Since this investigation was begun on the ionic materials, we begin with the simple Born potential:

$$V(R) = -A/R + B/R^m$$

which was the first used to characterize the cohesive energy of ionic crystals. This potential is the sum of an attractive Coulomb potential and a repulsive potential. At the equilibrium distance:

$$V(r_n) = -\frac{A}{r_n}(1-m^{-1})$$

and the potential is completely characterized by a constant A which includes the Madelung constant and the products of the atomic valences, the equilibrium distance, and the repulsive exponent m. The advantage of using this simple potential is its simplicity and the fact that, as we will see later, it yields good results.

To make the connection between this potential and the Morse potential, we must first recognize that the cohesive energy is far greater than the thermal energy of the crystal, since the former represents the energy necessary to separate all the ions by an infinitely large distance. Based on available empirical information concerning the thermal energy of the alkali halides and other ionic materials, we set the depth of the Morse potential as 10% of the cohesive energy, while the inverse width of the Morse potential (or more precisely ar_n) is determined by comparing the ratio of the cubic to the harmonic force constants for the Morse and Born potentials to yield the result $ar_n = (m+4)/5$.

These two empirical rules:

$$D = 0.1 \ V(r_n)$$

$$ar_n = (m+4)/5$$

allow the thermal expansion to be completely characterized by knowing the crystal structure, the Debye temperature, and the volume compressibility, from which the repulsive exponent is determined.

THERMAL EXPANSION IN ALKALI HALIDE AND HIGH
TEMPERATURE CRYSTALS

The first group of materials to which this procedure was applied was the alkali halide crystals. Because the relevant experimental data needed for the calculations have been available for these crystals for a long time, and the Born potential for these materials was characterized many years ago, the calculations for these materials was quite straight forward. However, it should be recognized that this use of the Born potential in determining the thermal expansion is entirely new. Table 1 tabulates the results of the calculations and compares them with experiment. It can be seen from the table that the results are quite good, considering the fact that only the crystal structure, Debye temperature, and volume compressibililty were used in the calculation with no adjustable parameters. The probable error in the calculations appears to be close to 10%.

Table 1.   Calculated and Observed Average Coefficients
           of Thermal Expansion (100°K - 600°K) for
           Various Alkali Halide Crystals.

	$\bar{\alpha}_{calc}(\times 10^6 \ °K^{-1})$	$\bar{\alpha}_{exp}(\times 10^6 \ °K^{-1})$
LiF	32.1	33.6
LiCl	39.1	44.9
NaF	36.0	33.2
NaCl	41.7	40.7
NaBr	43.3	42.7
KF	43.3	33.1
KCl	43.6	37.7
KBr	47.7	39.6

The next step in applying this model involves considering materials which display varying degrees of ionicity. Although the Born potential was developed for use with ionic crystals, the method which has been used here for characterizing the bottom of the potential well was tested for materials of varying crystal structures and ionicities. The results for a group of high temperature materials is presented in Table 2, in which the ionicity of the compounds listed decreases in descending order (MgO being the most ionic and diamond the least). The agreement between experimental and calculated values is about the same for all materials regardless of the degree of ionicity and about as good as for the alkali halides. These results indicate that this method of calculating the thermal expansion of insulators is applicable to a wide range of materials and not just to ionic materials. The agreement for polycrystalline samples of the noncubic crystals Al_2O_3 and TiO_2 indicates that the calculated results are not seriously affected by crystal anisotropy.

Table 2. Calculated and Observed Average Coefficients of Thermal Expansion (100 °K - 1400 °K) for Various High Temperature Materials.

	$\bar{\alpha}_{calc}(x10^6 \ °K^{-1})$	$\bar{\alpha}_{exp}(x10^6 \ °K^{-1})$
MgO	14.7	15.5
Al_2O_3*	10.1	9.4
TiO_2*	7.5	8.8
SiC	4.7	5.4
Diamond	4.0	4.1

*Polycrystalline Samples

Figure 1 compares the theoretical and experimental coefficient of thermal expansion curves for MgO. As discussed above, this model gives a simple connection between the shape of the coefficient of thermal expansion curve and the parameters of the Morse potential. At high temperatures, the leading contribution (dotted line) is nearly constant and has the overall shape of the Debye specific heat. The second order term is a fractional additive correction to the leading term which at high temperatures is proportional to kT/D. Thus the theoretical curve (solid line) is the sum of the leading term plus the correction. The magnitude of the correction at a given temperature thus fixes (at high temperatures) the value of D. Since the magnitude of

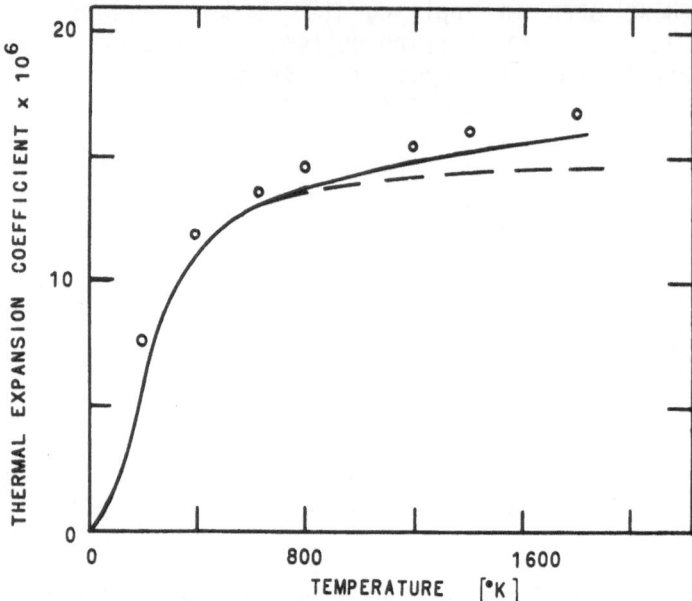

Fig. 1 – Calculated (solid line) and observed (open circles) values of
the coefficient of thermal expansion in MgO. The dashed line is the
contribution of the leading term

the leading term (at high temperatures) is inversely proportional to ar_nD, the value of the other parameter   a   is fixed
unequivocally.

CALCULATION OF THERMAL EXPANSION WITH NO
EXPERIMENTAL INPUT

In the calculations presented so far, the experimental
values of the Debye temperatures have been used.   However, the
Debye temperatures for many materials have not been determined
experimentally.   Consequently, it would be quite convenient if
the Debye temperature for each material could be approximately
determined from the potential well parameters.   This would by-
pass the need for the experimental values and at the same time
provide a check on the values of the potential well parameters
which have been calculated.

The Debye temperature corresponds to a maximum or cutoff
frequency associated with the normal modes.   A well-known result
for the monatomic linear chain is that the cutoff frequency is
twice the natural frequency of oscillation for one mass point
in the chain.   Using the parameters of the Morse potential, this

frequency is $2a(2D/m)^{1/2}$. Although the situation in a three dimensional crystal is not the same, the Debye temperatures for simple metals were calculated with good accuracy using this method. Similar calculations have been performed for the binary crystals considered so far with the anion-cation reduced mass used in the formula instead of the atomic mass. The calculated values once again agree rather well with experiment in these materials, indicating that the parameters of the Morse potential chosen by the method we have presented gives reasonable values, independent of bond type.

Ideally, it is desirable to estimate the thermal expansion of materials with no input other than the chemical formula of each compound. This is possible, with some penalty in accuracy, by using the approach which we have discussed. First, the simple empirical scaling rule:

$$m_Q - 1 = (m - 1)/Q$$

can be used to estimate the repulsive exponent for any cation-anion pair. This rule, which has been found to be applicable to a wide range of materials, relates the exponents for multivalent atoms having average valence Q to the known exponents for monovalent ions determined for the alkali halides which have the same electronic configuration. In addition, the Debye temperature can be estimated with some accuracy by means of the procedure just discussed, while the nearest-neighbor distances can be estimated quite accurately by means of Slater's table of atomic radii.[2] This leaves only the Madelung constant as an unknown. The empirical work of Kapustinsky[3] and Templeton[4] demonstrated that the Madelung constant can be approximated fairly accurately for any structure by means of a simple formula in which the result is proportional to the sum of squares of the atomic valences. If no structural information is used (i.e., only the chemical formula), then the calculated Madelung constant has a maximum error of about +10%. Thus, by using no information other than the chemical formula of the compound, it is possible to estimate the thermal expansion of any insultator, although the probable error in the calculation is increased over the values obtained when experimental information is used.

Table 3 summarizes the results of such calculations as applied to a group of binary and higher order materials. The calculated values use no experimental input and no adjustable parameters. Considering this fact, the accuracy of the calculations is quite good, and appears to be independent of crystal complexity and anisotropy.

The results presented here strongly suggest that the semi-

Table 3. Calculated and Observed Average Coefficients
of Thermal Expansion (300 °K - 1000 °K) for
Some Binary and Higher Order Compounds.

	$\bar{\alpha}_{calc}(\times 10^6\ °K^{-1})$	$\bar{\alpha}_{exp}(\times 10^6\ °K^{-1})$
CaO	16.0	12.4
UP	8.8	8.7
UN	8.0	8.9
BN	4.5	3.9
WC*	3.9	4.4
$BaTi_4O_9$	8.1	8.7
Mg_2SiO_4	9.1	10.9
$SrTiO_3$	9.9	11.1
Al_2TiO_5*	8.0	8.9

*Polycrystalline Samples

empirical method of calculating thermal expansion which has been
presented can predict the thermal expansion of a wide range of
insultating and ceramic materials with little or no experimental
input and no adjustable parameters. In some cases, it has been
found that some structural information is needed to improve the
calculations since the local symmetry is different from that in
"normal" cases and this significantly alters the thermal expansion. In rarer cases (a few percent of the total), the normal
mode behavior of some materials is such that a strongly negative
coefficient of thermal expansion occurs at low temperatures, and
this significantly alters the high temperature behavior. A
localized continuum model such as this cannot in its present form
account for this behavior. However, based on present experience,
this approach gives good estimates for the thermal expansion in
about 90% of the materials considered, with a good indication for
the other 10% of what the source of error is.

REFERENCES

1. A. R. Ruffa, Phys. Rev. B16, 2504 (1977).
2. J. C. Slater, J. Chem. Phys. 41, 3199 (1964).
3. A. Kapustinsky, Z. Phys. Chem. B22, 257 (1933).
4. D. H. Templeton, J. Chem. Phys. 23, 1826 (1955).

PRESSURE AND TEMPERATURE BEHAVIOR OF FRAMEWORK SILICATES AND NITRIDES

Louis Cartz

Department of Mechanical Engineering
Marquette University
Milwaukee, WI 53233

J.D. Jorgensen

Solid State Science Division
Argonne National Laboratory
Argonne, IL 60439

ABSTRACT

The response of silicates and nitrides to changes of pressure or of temperature depends on the flexibility of the bond angles of the crystal structures. Where bond angles can vary between neighboring tetrahedra linked at one corner, relative tilting or rotations occur which modify the three dimensional framework and this dominates the response of the crystal structure to temperature or pressure changes. Direct evidence for this behavior is available from measurements by time-of-flight neutron diffraction with multi-component profile refinement procedures, where atomic position changes have been determined to pressures of ~3 x 10^9 Pa. The importance of the framework structural type has been illustrated by studies of the flexible α-quartz structures of SiO_2 and GeO_2 which have volume compressibility coefficients $K_V \sim 26 \times 10^{-12} Pa^{-1}$, the partially flexible structures Si_2N_2O and Ge_2N_2O with $K \sim 9 \times 10^{-12} Pa^{-1}$, and the rigid structures α and β Si_3N_4 with $K_V \sim 3 \times 10^{-12} Pa^{-1}$.

The response of silicates to high pressures is different for the three-dimensional frameworks ($K_V \sim 22 \times 10^{-12} Pa^{-1}$), for chain-like pyroxenes ($K_V \sim 9 \times 10^{-12} Pa^{-1}$) and for the discrete silicate structures ($K_V \sim 7 \times 10^{-12} Pa^{-1}$). This suggests that their behavior corresponds to flexible, partially flexible and rigid frame-

147

work structure respectively. The silicate chains of all the pyroxenes respond to pressure in the same way with the same compressibility coefficient along the chain direction.

There is a less systematic variation of the thermal expansion
coefficients with structural type since the thermal vibrations can
modify the behavior of the framework structure. Information about
the response of silicate or nitride frameworks to stresses is more
easily and directly obtainable by high pressure studies which can
provide essential preliminary data towards understanding thermal
expansion characteristics.

INTRODUCTION

The thermal expansion behavior of framework crystal structures has been discussed extensively by Megaw (1,2,3) and by
Taylor (4,5,6) in terms of the rearrangements of the constituent
parts of the linked coordination polyhedra which can result in
dimensional differences greater than those due to bond length expansions. Megaw (1,2,3) has shown this to be the case by con-
trasting structures where the tilting of neighboring polyhedra is
permitted, with otherstructures where tilting does not occur. All
of these discussions with relation to temperature effects apply
equally to the effects of high pressure on crystal framework structures (7). With the application of pressure, neighboring polyhedra tilt relative to one another if the framework arrangement
and the bonding permits, and this has a marked effect on the compressibility coefficients of the crystal structure. This has been
shown in a study at pressures up to 3×10^9Pa of a group of silicate and nitride structures. The structures studied were flex-
ible frameworks (α-quartz SiO_2 and α-quartz GeO_2), partially flexible oxynitrides (Si_2N_2O and Ge_2N_2O) and rigid frameworks (α and β
Si_3N_4) (7). The observations have confirmed the interpretations
of Megaw (1,2,3) and of Taylor (4,5,6) on polyhedral tilting and
also provided quantitative measurements.

The experimental data have been taken using time-of-flight neutron diffraction with powder samples in a hydrostatic, supported
piston-cylinder cell (8,9). The details of the individual studies
of α and β Si_3N_4 (10,11), Si_2N_2O (12), Ge_2N_2O (13) and the α-quartz
forms of SiO_2 and GeO_2 (8) are published elsewhere. The relative
unit cell volumes versus pressure have been determined as well as
the effect of pressure on bond angles and bond lengths (7).

It has been clearly demonstrated that there are three distinct branches in the compressibility behavior dictated by the
structural type rather than the elements present.

The dominant effect of pressure on α-quartz SiO_2 and GeO_2 is

to reduce dramatically the Si-O-Si or Ge-O-Ge angle linking neighboring tetrahedra (7), with the bond lengths remaining essentially
constant over the range of pressure studied. This results in a
relatively large volume compressibility coefficient $K_v \sim 26 \times 10^{-12}Pa^{-1}$. The change with pressure in the Si-O-Si or Ge-O-Ge
bond angles between tetrahedra is again the dominant feature ob-
served for the oxynitrides Si_2N_2O and Ge_2N_2O with $K_v \sim 9 \times 10^{-12}Pa^{-1}$
while α and β Si_3N_4 exhibit no observable pressure effects on bond
angles or bond lengths and have a relatively small $K_v \sim 3 \times 10^{-12}Pa^{-1}$.
It has been concluded that the bonding arrangement of a framework
structure is the dominant factor in determining the compressibility
behavior (7).

        Since this should be a general property of crystal structures,
the volume compressibility coefficients of different silicate
structural types have been tabulated in Table 1. The three dimen-
sional networks have values of K_v ranging from 192 to 271 ($\times 10^{-13}Pa^{-1}$) and can be considered flexible framework structures.
The single-chain pyroxenes have values of K_v ranging from 70 to
107 ($\times 10^{-13}Pa^{-1}$) and are partially flexible. Structures with
discrete SiO_4 tetrahedra or 6-membered Si_6O_{18} rings range from
54 to 91 ($\times 10^{-13}Pa^{-1}$) corresponding to relatively rigid structures. There appears, once again, to be a relationship between
structural types and their overall behavior at high pressures.

        The pyroxene silicate structures are the largest group of
closely related silicates here considered and K_v varies from
70 to 107 ($\times 10^{-13}Pa^{-1}$), showing a steady variation with the
sizes of cations present. K_v is largest for diopside where the
Ca and Mg average ion size is 0.92A, and smallest for spodumene
where the average cationic radius is 0.68A. There is also a relationship between the cationic sizes of these monoclinic pyroxenes and their β-angle (14), so that the β-angles, cationic sizes
and K_v vary together (Table 2). A simple calculation can estimate
the anisotropic compressibility coefficients for the pyroxenes,
since K_v appears to depend on the "flexibility" of the chain of
silicate tetrahedra and on the size of the cations interspersed
between the chains. Assuming the compressibility coefficient
along the chain direction K_c for all the pyroxenes to be the same,
the coefficient perpendicular to the chains can be taken to be
$K_{\perp} = kr$ where k is a constant and r the cation radius, so that
$K_v = K_c + 2 kr$. Using the data in Table 2, k is found to be
78 $\times 10^{-13}Pa^{-1}A^{-1}$ and $K_c = -35 \times 10^{-13}Pa^{-1}$ for all the pyroxenes.
Experimental and calculated K_v are compared in Table 2 and seem
to be in good agreement. This supports the contention that the
variation with pressure of the Si-O-Si angle of the silicate chain
is the same for all the pyroxenes and indicates how the physical
properties of the silicates are determined by the nature of the
silicate framework, here a single chain of silicate tetrahedra.

TABLE 1.　Volume Compressibility Coefficients at Atmospheric Pressure.　(Ref. 16)

	Formula	Structural Type	$K_v (10^{-13} Pa^{-1})$
α-Quartz	SiO_2	Three-dimensional Network	271
α-Quartz	GeO_2	Three-dimensional Network	256
Nepheline	$Na\ Al\ SiO_4$	Tridymite-type Framework	205
Analcite	$Na\ Al\ Si_2O_6 \cdot H_2O$	Zeolite	197–370
Orthoclase	$K\ Al\ Si_3O_8$	Feldspar Framework	212
Albite	$Na\ Al\ Si_3O_8$	Feldspar Framework	202
Microcline	$K\ Al\ Si_3O_8$	Feldspar Framework	192
Diopside	$Ca\ Mg\ Si_2O_6$	Pyroxene (Single Chain)	107
Augite	$Ca(Mg,Fe,Al)(Al_2Si)_2O_6$	Pryoxene (Single Chain)	102
Hypersthene	$(Mg,Fe)\ SiO_3$	Pyroxene (Single Chain)	100
Aegirite	$NaFe\ Si_2O_6$	Pyroxene (Single Chain)	94
Jadeite	$NaAlSi_2O_6$	Pyroxene (Single Chain)	75
Spodumene	$Li\ AlSi_2O_6$	Pyroxene (Single Chain)	70
Tourmaline	$WX_3B_3Al_3(AlSi_2O_9)_3(O,OH,F)_4$	6-membered Si_6O_{18} Rings	81
Beryl	$Be_3Al_2Si_6O_{18}$	6-membered Si_6O_{18} Rings	54
Fayalite	Fe_2SiO_4	Discrete SiO_4	91
Zircon	$Zr\ SiO_4$	Discrete SiO_4	86
Staurolite	$FeAl_5Si_2O_{12}(OH)$	Discrete SiO_4	80
Forsterite	Mg_2SiO_4 (olivine, peridot)	Discrete SiO_4	80
Topaz	$Al_2SiO_4(F,OH)_2$	Discrete SiO_4	61
Garnet	$M_3N_2(SiO_4)_3$ M = Mg,Fe,Mn,Ca N = Al,Fe,Cr	Discrete SiO_4	55–67

TABLE 2. Cationic Sizes, Volume Compressibility, Thermal Expansion and the β Angle for the Monoclinic Pyroxene Structures. $K_V = K_C + 2kr$, where r is cationic radius in Å and $k = 78 \times 10^{-13}$Å^{-1}Pa^{-1} for pyroxenes, and $K_\perp = kr$. K_C is -35×10^{-13}Pa^{-1} for all the pyroxenes.

Formula	Average Cationic Size (Å)	β	α_V (10^{-6}°C^{-1}) Up to 400°C (Ref. 17)	K_V (10^{-13}Pa^{-1}) Measured (Ref. 16)	Calculated	$K_\perp$ (10^{-13}Pa^{-1}) Calculated
Diopside	CaMgSi$_2$O$_6$	0.92	105° 50'	28	107	109	72
Augite	Ca(Mg,Fe,Al)(Al$_2$Si)$_2$O$_6$	--	105°	25	102	-	--
Hypersthene	(MgFe)SiO$_3$	0.81	Orthorhombic	--	99-108	-	--
Aegirite	NaFeSi$_2$O$_6$	0.83	106° 49'	--	94	95	65
Jadeite	NaAlSi$_2$O$_6$	0.73	107° 26'	29	75	79	57
Spodumene	LiAlSi$_2$O$_6$	0.68	110° 20'	10	70	71	53

TABLE 3.   Volume thermal expansion coefficients (to 800°C) of
           silicates in terms of structural types; values for
           nitrides and oxynitrides given for comparison pur-
           poses. α_v for silicates from (17); for α-quartz
           SiO_2 and GeO_2 (8).

Structural Type		$\alpha_v (10^{-6} {}^\circ C^{-1})$
Framework	α-Quartz SiO_2	40^a
Framework	α-Quartz GeO_2	30^a
Framework	Nepheline (based on Trydimite)	49-72
Framework	Feldspars	20-33
Double Chain	Hornblende (Amphibole)	33
Single Chain	Pyroxenes	32-38
Discrete Si_2O_7	Akermanite (Melilite)	33
	Gehlenite (Melilite)	26
Discrete SiO_4	Olivine	31-44
	Merwinite (Chrysolite, Olivine)	42
	Garnets	26-30
	Andalusite	43
	Sillimanite	26
	Topaz	25
	Zircon	18

^{a}to 400°C

The volume thermal expansion coefficients (α_v) of silicates are listed in Table 3 by structural types including three dimensional framework structures (α-quartz SiO_2 and GeO_2, nepheline, feldspar), single-chain pyroxenes and double chain amphiboles, and structures with discrete Si_2O_7 and SiO_4 polyhedra. However, the values of α_v do not group as precisely as for K_v. The values of α_v are much more variable apparently because the effect of increased thermal vibrations modifies the behavior of the framework structure, whereas at increased pressures, the thermal vibration amplitudes are probably restrained so that the behavior due to framework type is accentuated.

CONCLUSIONS

The compressibility behavior of crystals is dominated by their atomic structural arrangement. Variable bond angles between polyhedra joined at a corner can lead to flexible type framework structures. It is these bond angle changes which dominate the overall behavior. Where bond angles are restrained by symmetry or coordination a much more rigid structure arises with quite different physical properties.

The thermal expansion behavior of crystals follow the same general principles but with many more variations. The effect of increased amplitudes of thermal vibration may well play an important role, so that the situation with increase of temperature is not as precisely defined as for high pressure effects. It should be very profitable, therefore, to carry out studies at high pressure of the behavior of the framework crystal structures as a preliminary to studies of the thermal expansion behavior.

ACKNOWLEDGMENTS

This work was supported by the U.S. Department of Energy, and National Science Foundation Grant No. DMR 76-05902.

REFERENCES

1. H.D. Megaw, "Crystal Structures; A Working Approach" (Saunders 1973), Chapter 14.
2. H.D. Megaw, Mat. Res. Bull. 6, 1007, (1971).
3. H.D. Megaw, Acta Cryst. A24, 589, (1968).
4. D. Taylor, Min. Mag. 38, 593, (1972)
5. D. Taylor, Min. Mag. 36, 761, (1968)
6. D. Taylor and C.M.B. Henderson, Amer. Min. 53, 1476, (1968).
7. L. Cartz and J.D. Jorgensen, J. Appl. Phys. (1980) (in press).
8. J.D. Jorgensen, J. Appl. Phys. 49, 5473, (1978).
9. T.G. Worlton, J.D. Jorgensen, R.A. Beyerlein, and D.L. Decker, Nuclear Instrum. Methods 137, 331 (1976).

10. S.R. Srinivasa, "High Pressure Neutron Diffraction Study of Ceramic Nitrides," Ph.D. Thesis, Marquette University, Milwaukee, WI, (1977).

11. J.D. Jorgensen, T.G. Worlton, S.R. Srinivasa, L. Cartz, Proc. Conf. Neutron Scattering, Gatlinburg, June 1976, p. 55 (ORNL and USERDA).

12. S.R. Srinivasa, L. Cartz, J.D. Jorgensen, T.G. Worlton, R.A. Beyerlein and M. Billy, J. Appl. Cryst. 10, 167 (1977).

13. S.R. Srinivasa, L. Cartz, J.D. Jorgensen, and J.C. Labbe, J. Appl. Cryst., 12, 511 (1979).

14. E.J.W. Whittaker, Acta Cryst., 13, 741, (1960).

15. R.W.G. Wyckoff, "Crystal Structures," 2nd Edition, (Interscience 1968).

16. S.P. Clark Jr. (Editor), "Handbook of Physical Constants" (Revised Edition 1966), Geol. Soc. of America, Memoir 97.

17. R.H. Stutzman, J.R. Salvaggi, and H.P. Kirchner, Cornell Aeronautical Lab., Inc. Report P1-1273-M-4, (1959).

ITES SESSION 6: MISCELLANEOUS MATERIALS/APPLICATIONS

Session Chairman: W. A. Plummer
                     Corning Glass Works
                     Painted Post, NY

HIGH PRESSURE AND HIGH TEMPERATURE STUDIES ON MERCUROUS CHLORIDE

Y. C. Venudhar, T. Ranga Prasad, Leela Iyengar,
K. Satyanarayana Murthy, and K. V. Krishna Rao

Department of Physics, University College of Science
Osmania University
Hyderabad-500 007, India

INTRODUCTION

Crystals of mercurous chloride (Hg_2Cl_2) possess a number of unusual and interesting physical properties.[1] The crystal belongs to the space group I4/mmm (D_{4h}^{17}).[2] The structure contains chains of linear Cl-Hg-Hg-Cl molecules parallel to the c-direction. The centers of the molecules of the nearest neighbor chains are displaced c/2 along c. The structure transforms to a phase of lower symmetry at low temperature whose structure is uncertain.[3,4] Richter et al.[5] studied the Raman spectra of mercurous chloride at room temperature and at pressures up to 16 kbar. They report that the changes in the Raman spectra may be explained on the basis of the change of structure similar to the structure changes at low temperatures. Rosasco et al.[1] found by X-ray studies that at low temperatures there is essentially no change in the value of 'c', whereas 'a' reduces considerably. Hence it is thought desirable to undertake X-ray studies on mercurous chloride at high pressures and high temperatures. This paper gives an account of the preliminary results obtained.

HIGH PRESSURE STUDIES

X-ray diffraction data of Hg_2Cl_2 at room temperature and at different pressures have been obtained using a diamond anvil apparatus (XKB-100) supplied by the Materials Research Corporation of the United States. MoK_α radiation with Zr filter was used from a normal focus X-ray tube. Hg_2Cl_2 of 99% purity supplied by Sarabhai Chemicals of Baroda was used in this study. The lattice parameters at 30°C were determined using a Unicam 19 cm high temperature powder

camera and FeK$_\alpha$ radiation. The lattice parameters are a = 4.4850 ±
0.0001 Å, and c = 10.955 ± 0.001 Å, in agreement with the values
reported by earlier workers.[4] The change in diffraction pattern
started at a pressure of about 5 kbar and ended at about 20 kbar.
The pattern recorded at 20 kbar could be indexed, based on an ortho-
rhombic lattice with lattice parameters a = 4.23 Å, b = 4.54 Å, and
c = 10.44 Å. Further details of the results are given in a recent
publication by Ranga Prasad et al.[6]

HIGH TEMPERATURE STUDIES

The diffraction patterns of Hg_2Cl_2 have been recorded at different temperatures, ranging from room temperature (30°C) to 260°C,
using a Unicam 19 cm high temperature powder camera and FeK$_\alpha$ radiation. The details of the experimental arrangement and the evalua-
tion of the precise lattice parameters and the coefficients of
thermal expansion have been given in earlier papers by Krishna Rao
et al.[7,8] The specimen for the study was prepared by filling the
powder in a thin-walled quartz capillary. The diffraction lines
314, 118, 208, and 404 recorded in the Bragg angle region 50° to
70° have been used for determining the lattice parameters.

The lattice parameters and the volume of the unit cell at different temperatures are given in Table 1 and are shown graphically
in Figures 1 and 2. While 'a' parameter increases with temperature,
'c' parameter and the volume of the unit cell decrease with increasing temperature. The coefficient of thermal expansion along the
c-axis, i.e., $\alpha_{||}$ is found to have a constant value, -31.94×10^{-6}
°C^{-1}, throughout the range of temperatures studied. Table 2 gives
$\alpha_\perp$ and β, the volume coefficient of expansion at different temperatures. They can be represented by Equations 1 and 2:

$$\alpha_\perp = 8.3370 \times 10^{-6} + 2.9494 \times 10^{-8}T \tag{1}$$

$$\beta = -15.2624 \times 10^{-6} + 5.8991 \times 10^{-8}T \tag{2}$$

where T is temperature in °C. The values calculated from these
equations are also given in Table 2 for comparison.

DISCUSSION

Though many crystals exhibit negative thermal expansion along
one axis, their coefficients of volume expansion are usually positive, especially at high temperatures. Some crystals having diamond
structure like silicon and germanium show negative volume expansion
at low temperatures.[9,10] So far only two structures β-AgI and
thorium fluoride were reported to have negative volume expansion
at high temperatures.[11,12] Thus it is unusual for Hg_2Cl_2 to show
at high temperatures a large negative expansion along the c-axis

TABLE 1. Lattice Parameters and Volume of Unit Cell
of Hg_2Cl_2 at Different Temperatures

Temperature, °C	a, Å	c, Å	V, Å^3
30	4.4850	10.955	220.36
94	4.4874	10.939	220.28
126	4.4896	10.921	220.13
160	4.4912	10.908	220.02
192	4.4934	10.894	219.96
260	4.4976	10.871	219.90

Table 2. Coefficients of Thermal Expansion ($\alpha_\perp$) and Volume
Expansion (β) of Hg_2Cl_2 at Various Temperatures

Temperature, °C	$\alpha_\perp \times 10^6$/°C		$\beta \times 10^6$/°C	
	Observed	Calculated	Observed	Calculated
40	9.48	9.52	-12.98	-12.90
60	10.15	10.11	-11.64	-11.72
80	10.70	10.70	-10.54	-10.54
100	11.26	11.29	- 9.42	- 9.36
120	11.93	11.88	- 8.08	- 8.18
140	12.38	12.47	- 7.18	- 7.00
160	13.16	13.06	- 5.62	- 5.82
180	13.60	13.65	- 4.74	- 4.64
200	14.27	14.24	- 3.40	- 3.46
220	14.83	14.83	- 2.28	- 2.28
240	15.39	15.42	- 1.16	- 1.10
260	--	16.01	--	- 0.08

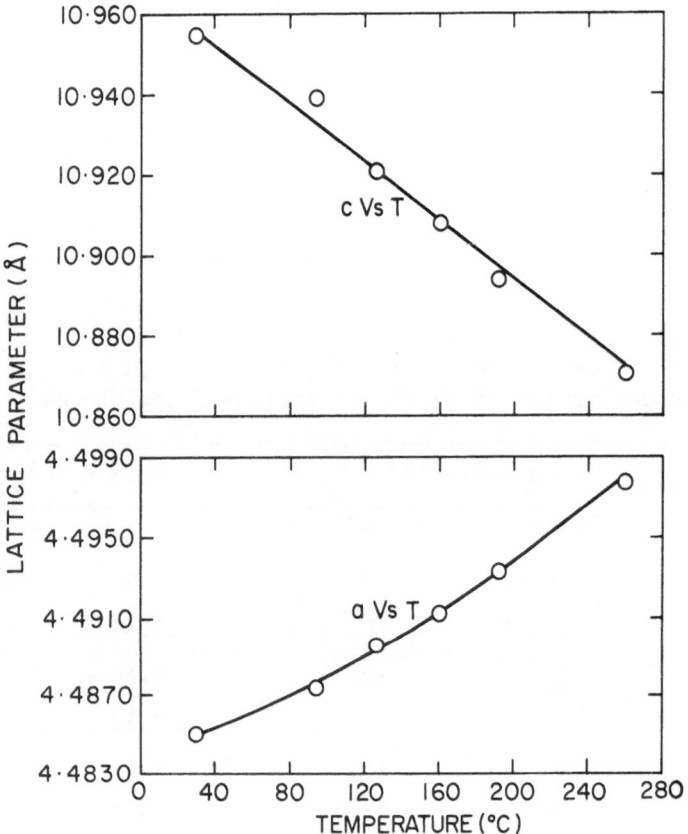

Figure 1.   Variation of lattice parameters of
            Hg_2Cl_2 with temperature.

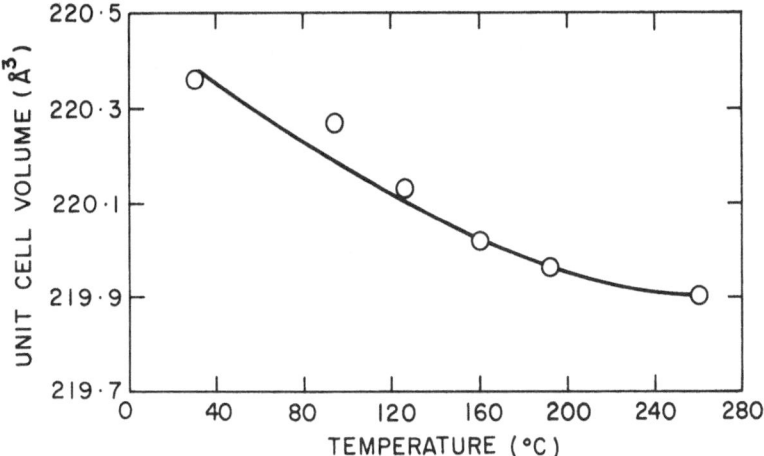

Figure 2.   Plot of unit cell volume of Hg_2Cl_2
versus temperature.

and also an overall negative volume expansion. An attempt to explain this unusual behavior of Hg_2Cl_2 in terms of the bonding in
different directions is under way.

ACKNOWLEDGMENTS

The authors are grateful to the University Grants Commission,
New Delhi, India for financing a research scheme under which the
work was done. One of the authors (Venudhar) desires to express
his gratitude to the Council of Scientific and Industrial Research,
New Delhi, India for awarding him a Senior Research Fellowship.

REFERENCES

1.   G. J. Rosasco, H. S. Parker, R. S. Roth, R. A. Forman and
      W. S. Brower, J. Phys. C: Solid State Phys., 11:35 (1978).
2.   R. W. G. Wycoff, Crystal Structure, Interscience, 2:36 (1948).
3.   C. Barta, A. A. Kaplyanskii, V. V. Kulakov and Y. F. Markhov,
      JETP Lett., 21:54 (1975).
4.   M. E. Boiko and A. A. Voilpolin, Sov. Phys. Solid State,
      19:1117 (1977).
5.   P. W. Richter, P. T. T. Wong and E. Whalley, J. Chem. Phys.,
      67:2348 (1977).
6.   T. Ranga Prasad, K. Satyanarayana Murthy, L. Iyengar and
      K. V. Krishna Rao, Pramana, 12:523 (1979).
7.   K. V. Krishna Rao S. V. Nagender Naidu and L. Iyengar,
      J. Appl. Cryst., 6:136 (1973).

8.   K. V. Krishna Rao, S. V. Nagender Naidu and P. L. N. Setty,
      Acta Cryst., 15:528 (1962).
9.   P. W. Sparks and C. A. Swenson, Phys. Rev., 163:779 (1967).
10. R. H. Carr, R. D. McCammon and G. K. White, Phil. Mag.,
      12:157 (1965).
11. B. R. Lawn, Acta Crysta., 17:1341 (1964).
12. R. L. Barns, Mat. Res. Bull., 12:327 (1977).

ENHANCEMENT OF THERMAL EXPANSION ANOMALY IN Fe-B AMORPHOUS INVAR

ALLOYS BY COLD ROLLING

K. Fukamichi*, H. M. Kimura, M. Kikuchi[+], and T. Masumoto

The Research Institute for Iron, Steel and Other Metals
Tohoku University, Sendai, 980, Japan

ABSTRACT

Fe-B amorphous alloys were prepared from the melt by rapid quenching method. These specimens were annealed or cold-rolled. The saturation magnetization and the Curie temperature were almost independent of the ratio of cold rolling. However, the anomaly in the thermal expansion below the Curie temperature of $Fe_{83}B_{17}$ amorphous alloy was remarkably enhanced by cold rolling, accompanying a negative thermal expansion in a wide temperature range. This enhancement was pronounced with the increase in the ratio of cold rolling. Such thermal expansion behaviors are very similar to those of Fe-Ni crystalline Invar alloys. The crystallized specimen showed no Invar characteristics, indicating a large thermal expansion such as that of normal metals and alloys.

INTRODUCTION

The magnetic and elastic properties of amorphous alloys have been studied extensively in recent years[1~2], whereas the thermal expansion data of them are not so sufficient. We have already measured the thermal expansion of several kinds of Fe-base and

*Present temporary address; IBM Thomas J. Watson Research Center, Yorktown Heights, NY 10598

[+]The Research Institute of Electric and Magnetic Alloys, Sendai, 982, Japan

Co-base amorphous alloys, and found that Fe-base amorphous alloys
show a remarkable anomaly below the Curie temperature[3]. As is
well known, such a large anomaly in the thermal expansion is caused
by the spontaneous volume magnetostriction. Recently, it has been
demonstrated that Fe-B amorphous alloys show the representative
Invar characteristics in a wide temperature range, accompanying a
pronounced thermal expansion anomaly[4,5]. The high-field susceptibility[6] and the forced volume magnetostriction[7] of Fe-B amorphous
alloys are very large and comparable to those of Fe-Ni crystalline
Invar alloys[8,9]. The Curie temperature decreases with increasing
Fe content, and the concentration dependence of the magnetic moment
per Fe atom shows a broad maximum around 15% B[5,9]. Almost all of
these behaviors are very similar to those of Fe-Ni crystalline
Invar alloys[10].

The thermal expansion characteristics of Fe-Ni crystalline
Invar alloys is easily affected by cold working[11~14], and the
thermal expansion coefficient depends on the rolling direction.
This phenomenon has been called the $\Delta\alpha$ effect[11]. It has been
reported that the magnetic properties of amorphous alloys are
sensitively affected by plastic deformation[15~17]. Since the nearest neighbor coordination number in amorphous alloys, in general,
is close to that of f.c.c. crystalline alloys[18], and plastic
deformation of amorphous alloys causes a slip band like mode[19,20]
which is very similar to that of crystalline alloys. In addition,
as mentioned above, the magentic properties of Fe-B amorphous alloys
are similar to those of Fe-Ni crystalline Invar alloys. Therefore,
it would be expected that the $\Delta\alpha$ effect is observed in Fe-B amorphous alloys. Then, it is interesting to investigate the effect of
cold rolling on the thermal expansion and on the magnetic properties of Fe-B amorphous alloys.

In the present paper, we report the thermal expansion behaviors of as-prepared, annealed, cold-rolled and crystallized specimens of Fe-B alloys, together with their magnetic properties.

EXPERIMENTAL

Fe-B amoprhous ribbon samples with different width and 20~30μ
in thickness were prepared from the melt by a rapid quenching   .
method. The heat-treatment was carried out at 200°C for 2 hr. The
cold rolling was made with a rolling speed of 15 rpm at room temperature by using rollers with 50 mmφ. The ratio of cold rolling was
obtained from the change in the length of specimen. In order to
obtain the saturation magnetization, the Curie temperature and the
crystallization temperature, the temperature dependence of magnetization was measured by a Farady method in the magnetic field of 10
KOe. The thermal expansion was measured by the thermal expansion

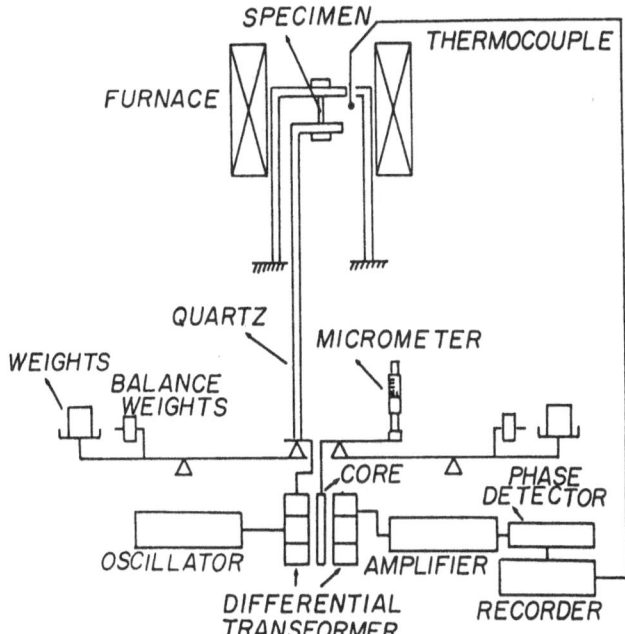

Fig. 1.   Schematic diagram of the thermal expansion
          apparatus

apparatus (Rigaku Thermoflex 8561 A1) shown in Fig. 1.   The
effective length of the sample is 10 mm.   The specimen is held
between the plates of the Inconel alloy.   The quartz rod for detection of the expansion is attached with the differential transformer
field, and the micrometer is also connected with the differential
transformer core.   By controlling the balance weight, we can give
a tension to the specimen.   A strong tension brings about a creep
elongation because the specimen is very thin, and in general, the
glass transition temperature of amorphous alloys is relatively low.
If a tension is not applied to the specimen, the thermal expansion
is not accurately detected because the specimen is not held in a
straight line.   Then, in the present study, 5 g was applied as a
dead weight.   The position of the core is adjusted by the micrometer.   The displacement is amplified and the output through the
phase detector is recorded.   In the present investigation, the
heating has been made by using the electric furnace in an argon
atmosphere with a rate of 2.5 °C/min.   The temperature was measured
with a platinel thermocouple.

RESULTS AND DISCUSSION

Figure 2 shows the saturation magnetization σ_S and the Curie temperature T_C of $Fe_{83}B_{17}$ amorphous alloy as a function of the ratio of cold rolling. The value of σ_S was extrapolated from the nitrogen temperature to 0 K following the $T^{3/2}$ law. The Curie temperature was determined from the temperature dependence of the magnetization. As shown in the figure, within the experimental error, these values are independent of the ratio of cold rolling. Luborsky et al. have obtained the similar result about the Curie temperature of $Fe_{40}Ni_{40}P_{14}B_6$ amorphous alloy[15]. In the case of Fe-Ni crystalline Invar alloys, the magnetic properties are influenced by cold working. For example, the saturation magnetization of (110) [001] rolling of $Fe_{71}Ni_{29}$ single crystal decreases and its high-field susceptibility increases with increasing the reduction of cold rolling[21]. This means that the Invar characteristics of Fe-Ni crystalline alloy is enhanced by cold rolling. On the other hand, as mentioned above, the Curie temperature and the saturation magnetization of $Fe_{83}B_{17}$ amorphous alloy does not change. In the case of amorphous alloys, it has been confirmed from the measurements of magnetic and elastic properties that the stress is relieved by heating at temperatures below their crystallization temperature[22,23]. The Curie temperature of Fe-B amorphous alloys is very close to the crystallization temperature[4]. Therefore, the stress

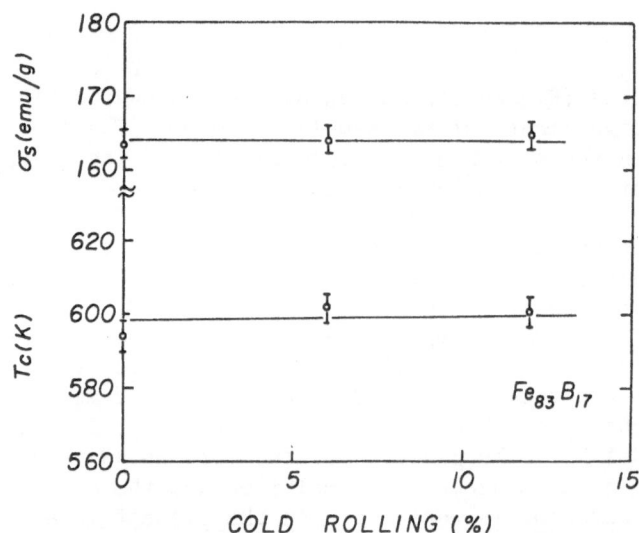

Fig. 2. Saturation magnetization and the Curie temperature of $Fe_{83}B_{17}$ amorphous alloy as a function of the ratio of cold rolling.

developed by cold rolling would be relieved during the measurement of the temperature dependence of the magnetization, resulting in no effect of cold rolling on the Curie temperature.

Figure 3 shows the thermal expansion curves of the as-prepared, annealed and cold rolled specimens. The rolling reduction of $Fe_{83}B_{17}$ amorphous alloy with about 0.4 mm in width is 5% and 12%. The thermal expansion was measured parallel to the rolling direction. The curve of the annealed specimen exhibits excellent Invar characteristics in a wide temperature range below the Curie temperature, and the thermal expansion coefficient α is almost zero at room temperature. Next, the expansion is very marked between the Curie temperature T_C and the crystallization temperature T_X, and then drastically shrinks above T_X. The value of α for the as-prepared specimen is almost same that of the annealed one at room

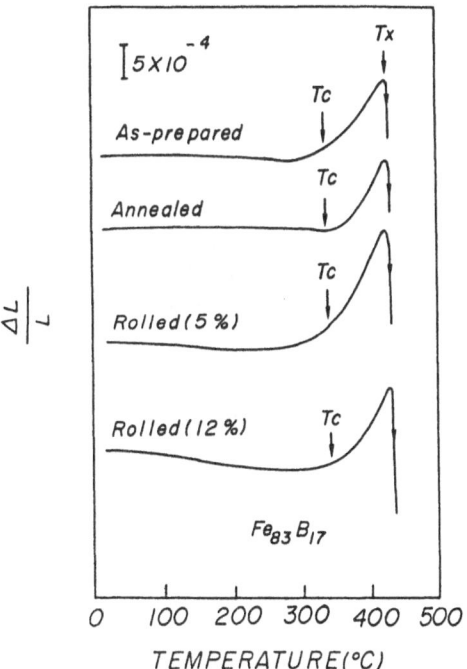

Fig. 3.   Thermal expansion curves of the as-prepared, annealed and cold rolled specimens of $Fe_{83}B_{17}$ amorphous alloy with about 0.4 mm in width and 20μ in thickness.

temperature, but the feature of curve around T_C is somewhat different. Such a smearing in the thermal expansion curve around T_C would be correlated to the structural relaxation[24], accompanying a release of stress induced by rapid quenching. On the other hand, the thermal expansion anomaly is significantly enhanced by cold rolling, and the curves exhibit a negative expansion coefficient in a wide temperature range. The enhancement of the thermal expansion anomaly for other Fe-B amorphous alloys with the different compositions has been also observed. This behavior is very similar to that of $Fe_{64}Ni_{36}$ crystalline Invar alloy[13]. The following experimental results would be interesting, that is, the randomness of amorphous alloy is increased by cold rolling, resulting in the decrease in the small angle X-ray scattering intensity[15], and the thermal expansion anomaly of Fe-B amorphous alloys is enhanced by cold rolling as mentioned above. On the other hand, the thermal expansion anomaly of Fe-Pt crystalline Invar alloys is more remarkable in the disordered state[25]. As seen in the figure, the crystallization temperature T_X of the cold-rolled specimens almost coincides with that of the annealed and the as-prepared specimens. Luborsky et al. also have investigated the effect of cold rolling on the crystallization temperature T_X of $Fe_{40}Ni_{40}P_{14}B_6$ amorphous alloy, and found that T_X is independent of cold rolling[15].

In Fig. 3, we have pointed out that the specimens drastically shrink around the crystallization temperature T_X. Figure 4 shows the thermal expansion curves of $Fe_{83}B_{17}$ alloys annealed at 200°C for 2 hr or crystallized above T_X. As described in Fig. 3, the annealed specimen shows the Invar characteristics in a wide temperature range. However, as seen in the figure, the crystallized specimen shows no Invar characteristics as indicated by a large thermal expansion such as that of normal metals and alloys. We have also measured the thermal expansions of FeB, Fe_2B and Fe_3B crystalline compounds, but no remarkable thermal anomaly has been observed. These facts mean that the Invar characteristics originate from the amorphous structure in Fe-B alloy.

In the case of Fe-Ni crystalline Invar alloys, the thermal expansion coefficient depends on the rolling direction, and this behavior was termed the $\Delta\alpha$ effect by Schulze[11]. The $\Delta\alpha$ effect is usually defined by; $\Delta\alpha = \alpha_{PD} - \alpha_{RD}$, where α_{PD} and α_{RD} are the thermal expansion coefficients measured perpendicular and parallel to the rolling direction, respectively. The values of $\Delta\alpha$ for Fe-Ni crystalline Invar alloys become positive at room temperature[12,13]. It would be expected that Fe-B amorphous Invar alloys also show the $\Delta\alpha$ effect because the thermal expansion characteristics are easily affected by cold rolling as shown in Fig. 3. In addition, the magnetic properties of Fe-B amorphous alloys are very similar to those of Fe-Ni crystalline Invar alloys[5]. In order to investigate the $\Delta\alpha$ effect, $Fe_{83}B_{17}$ amorphous alloy with about 10 mm in width

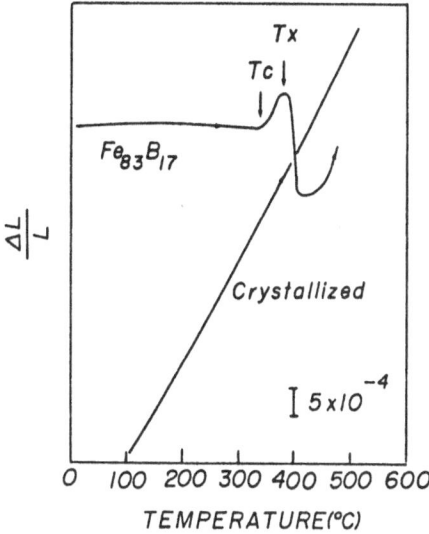

Fig. 4.   Thermal expansion curves of the annealed and
crystallized specimens of $Fe_{83}B_{17}$ alloy.

was cold-rolled. Unfortunately, reliable results of the thermal
expansion have not been obtained due to the following reasons. The
amorphous ribbon samples are originally very thin, and their thickness is not uniform. Generally, the center is thicker than the
edge of the specimens in the direction of width. Then, in the
present study, the uniform cold rolling from the edge to the center
has been severe. The representative thermal expansion curves of
$Fe_{83}B_{17}$ amorphous alloy is shown in Fig. 5. As seen in the inset,
the specimens were cut from the wide sample rolled 12% as a whole.
The thermal expansion of these three parts of (A), (B) and (C) was
measured parallel to the rolling direction. The enhancement of the
thermal expansion anomaly for (C) is more remarkable than that of
(A). From the observation of the cross section of wide specimen,
it was made sure that the part of (C) is thicker than the part of
(A) before the cold rolling, then the reduction of the former would
be larger than that of the latter. This presumption is consistent
with the results in Fig. 3 because the thermal expansion anomaly is
enhanced with an increase of the cold roll reduction. Tino and
Kobayashi have discussed the enhancement of the thermal expansion
induced by cold working. They have assumed that Invar alloys are
regarded as possessing a variety of structures between the γ-phase
and α-phase alloys, and pointed out that the shear mechanism of γ ⇄
α martensitic transformation plays important roles in the Invar
alloys[13]. However, in the case amorphous alloys needless to say,
they have no relationship with the martensitic transformation.

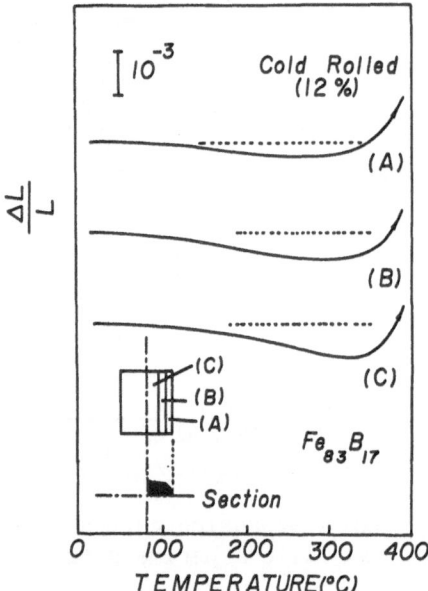

Fig. 5.   Thermal expansion curves of the parts of the cold
          rolled $Fe_{83}B_{17}$ amorphous wide specimens with
          about 10 mm in width.

Recently, Kagawa and Chikazumi have found that the variation of the
anisotropy in the thermal expansion of the cold rolled Fe-Ni Invar
alloy with an increase of the roll reduction is quite similar to
that of roll magnetic anisotropy[14]. They have pointed out that
Fe-Fe pairs are changed by cold rolling, resulting in such similar
behaviors[14]. Kanamori has discussed the Invar problems by using
the coherent potential approximation calculation, and he has
concluded that Fe-Fe pairs are responsible for the Invar effect[26].
Recently, we have found that the thermal expansion anomaly below
the Curie temperature slightly depends on the width of the as-
prepared amorphous ribbons[27]. It is suggestive that the magnetic
anisotropy is affected by the condition of sample preparation even
if the alloy composition is the same as each other. Then, the
investigation of the magnetic anisotropy of Fe-B amorphous alloys
would be very important to discuss in more detail the enhancement
of the thermal expansion anomaly.

SUMMARY

The magnetic properties of Fe-B amorphous Invar alloys are very similar to those of Fe-Ni crystalline Invar alloys, then, it would be expected that the effect of the cold rolling on the thermal expansion anomaly of Fe-B amorphous alloys is remarkable. We have investigated the magnetic properties and the thermal expansion characteristics of the as-prepared, annealed, cold-rolled and crystallized specimens, and the following results have been obtained;
(a) The saturation magnetization, the Curie temperature and the crystallization temperature are almost independent of the ratio of cold rolling.
(b) The thermal expansion anomaly below the Curie temperature is remarkably enhanced, resulting in a negative thermal expansion in a wide temperature range.
(c) This enhancement becomes more remarkable with an increase of the cold roll reduction.
(d) The thermal expansion curves of as-prepared and cold-rolled specimens are somewhat different from that of the annealed specimen around the Curie temperature.
(e) The crystallized specimen does not show the Invar character-istics.

ACKNOWLEDGMENTS

The authors are indebted to Dr. T. Takahashi of the Tohoku Metal Industries LTD. and Mr. K. Obara of the Research Institute for Iron, Steel and Other Metals, Tohoku University for their preparation of wide specimens and cold rolling. They are also grateful to Dr. H. Masumoto of the Research Institute of Electric and Magnetic Alloys for his continuing encouragements and supports during the present work.

REFERENCES

1. *AMORPHOUS MAGNETISM*, ed. H. O. Hooper and A. M. de Graff, Plenum Press, New York-London (1973).
2. *AMORPHOUS MAGNETISM II* , ed. R. A. Levy and R. Hasegawa, Plenum Press, New York (1977).
3. K. Fukamichi, M. Kikuchi and T. Masumoto, Sci. Rep. RITU, A26, 225(1977).
4. K. Fukamichi, M. Kikuchi, S. Arakawa and T. Masumoto, Solid State Commun., 23, 955(1977).
5. K. Fukamichi, T. Masumoto and M. Kikuchi, IEEE Trans. Magn. MAG-15, 1404(1979).
6. H. Hiroyoshi, K. Fukamichi, M. Kikuchi, A. Hoshi and

T. Masumoto, Phys. Lett., 65A, 163(1978).

7. K. Fukamichi, M. Kikuchi, S. Arakawa , T. Masumoto,
    T. Jagielinski, K. I. Arai and N. Tsuya, Solid State Commun.,
    27, 405(1978).

8. K. Fukamichi, M. Kikuchi, H. Hiroyoshi and T. Masumoto, Third
    Int. Conf. Rapidly Quenched Metals, Vol. 2, 117(1978).

9. K. Fukamichi, H. Hiroyoshi, M. Kikuchi and T. Masumoto, J.
    Magn. Magn. Mater., 10, 294(1979).

10. *PHYSICS AND APPLICATIONS OF INVAR ALLOYS*, ed. H. Saito, Maruzen
    Co. Ltd., (1978).

11. R. Schulze, Z. Angew. Phys., 7, 57(1955).

12. G. Pupke, Z. Phys. Chem., 207, 91(1957).

13. Y. Tino and M. Kobayashi, J. Phys. Soc. Japan, 41, 59(1975).

14. H. Kagawa and S. Chikazumi, J. Phys. Soc. Japan, 43, 1097
    (1977).

15. F. E. Luborsky, J. L. Walter and D. G. LeGrand, IEEE Trans.
    Magn. MAG-12, 930(1976).

16. M. R. J. Gibbs, J. P. Patterson and J. E. Evetts, Phys. Lett.,
    68A, 461(1978).

17. M. R. J. Gibbs, J. E. Evetts and N. J. Shah, J. Appl. Phys.,
    50, 7642(1979).

18. G. S. Cargill III, *Solid State Physics*, Vol. 30, 227(1975).

19. T. Masumoto and R. Maddin, Acta Met, 19, 725(1971).

20. H. S. Chen and D. E. Polk, J. Non-Cryst. Solids, 15, 174(1974).

21. S. Chikazumi, T. Mizoguchi, N. Yamaguchi and P. Beckwith,
    J. Appl. Phys., 39, 939(1968).

22. F. E. Luborsky, J. J. Becker and R. O. McCary, IEEE Trans.
    Magn. MAG-11, 1644(1975).

23. M. Kikuchi, K. Fukamichi and T. Masumoto, Sci. Rep. RITU, A26,
    232(1977).

24. H. S. Chen and E. Coleman, Appl. Phys. Lett., 28, 245(1976).

25. A. Kussmann, M. Auwärter and G. G. V. Rittberg, Ann. Phys.,
    6, No. 4, 174(1948).

26. J. Kanamori, *PHYSICS AND APPLICATIONS OF INVAR ALLOYS*, ed.
    H. Saito, Maruzen Co. Ltd. Tokyo (1978). 221.

27. K. Fukamichi, H. M. Kimura, M. Kikuchi and T. Masumoto, to be
    submitted.

THERMAL EXPANSION BEHAVIOR OF POROUS ROCKS UNDER STRESS

Roy F. Greenwald, Paul I. Ashqar, and Wilbur H. Somerton

University of California
Berkeley, California

ABSTRACT

The axial thermal expansions and pore volume contractions of water-saturated Berea, Bandera, and Boise sandstones under stress have been determined as functions of temperature. The samples were subjected to a constant confining stress of 20.7 MPa and pore pressure of 6.9 MPa while being heated from 35°C to 175°C. Resultant decreases in the porosities were 4.4, 3.8, and 3.1 percent for Bandera, Berea and Boise sandstones, respectively. The effect of a variation in stress on Boise sandstone was also investigated. It was found that by decreasing the confining stress to 13.8 MPa the porosity reduction was decreased to 2.6 percent.

INTRODUCTION

The thermal expansion behavior of fluid-saturated porous rocks under stress is a topic of concern in petroleum and geothermal reservoir engineering, underground disposal of nuclear wastes, and similar projects where rocks are heated. One of the more sophisticated oil recovery techniques is to reduce the viscosity of the oil by heating it with injected steam. Although the principle is simple, the effect that elevated temperature has on the volumetric response of fluid-saturated sandstones under stress has not been investigated. The present study was conducted to determine the response of both the bulk volume and the pore volume of reservoir-type rocks to being heated.

Three types of outcrop sandstones included in this work are Bandera, Berea, and Boise sandstones which have porosities of

173

approximately 20, 25 and 27 percent, respectively. The porosity and volumetric data for the particular cores used in the tests are listed in Table 1. The mineralogical compositions of all three sandstones as given by Mann and Fatt (1960) are very similar, with quartz being the major component; quartz content is 35 percent for Bandera, 40 percent for Boise, and approximately 65 percent for Berea.

Conducting experiments under stress on liquid-saturated sandstones requires controlling two separate stresses. A confining stress is imposed hydrostatically on a jacketed core sample while the pore stress, or pressure of the saturating fluid, is controlled independently. The difference between the confining stress and pore pressure is frequently referred to as the effective stress.

Table 1.   Porosity and Volumetric Data for Test Specimens

Core	Bulk Volume, cc	Pore Volume, cc	Porosity
Bandera 1	103.10	20.62	20.0
Berea 1	105.97	25.72	24.3
Boise 1	104.32	27.85	26.7
Boise 2	102.55	27.57	29.9

LITERATURE REVIEW

Very little work has been done to determine the bulk thermal expansion of liquid-saturated sandstones or to investigate thermal pore volume changes under elevated stress conditions. None of the studies reports the simultaneous determination of the thermal response of pore and bulk volumes of samples under stress.

Skinner (1966) lists thermal expansions from earlier works for several different geological materials in an unstressed state. His value for an average linear expansion coefficient in the temperature range of 20° to 200°C is $10 \pm 2 \times 10^{-6}/°C$. This agrees quite well with data obtained in the present work.

Somerton and Selim (1961) measured the thermal expansion of dry Bandera, Berea, and Boise sandstones in the temperature range of 25°C to 1000°C at atmospheric pressure. These results are shown in Figure 1 for these sandstones and pure quartz both parallel and perpendicular to the C axis. The close similarity in the expansion of the three sandstones to that of quartz led the investigators to propose that the presence of quartz above a certain amount controls the thermal expansion behavior of the sandstones.

The axial thermal expansion of porous limestones is given by Harvey (1967). He presents thermal expansion curves for over 100

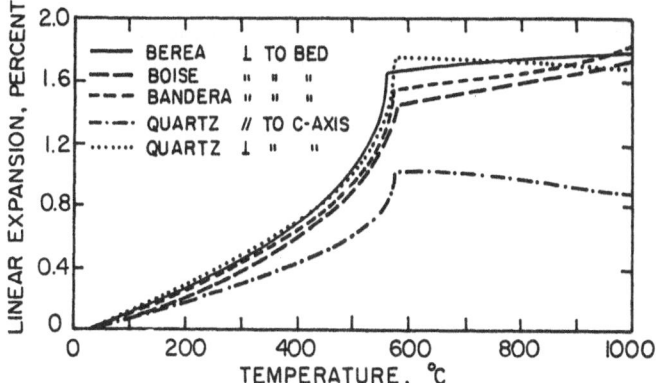

Figure 1.   Linear expansion of Bandera, Berea, Boise sandstones and
            quartz.   (After Somerton and Selim 1961)

calcitic and dolomitic dry limestone specimens. The temperature
range was -20°C to 80°C with many of the curves showing a discontinuity at approximately 20°C. Analysis of the strain-temperature
curves shows that the coarse-grained samples show a more linear
response. Values of the thermal expansion coefficient range from
as low as 3.4×10^{-6}/°C to as high as 9.9×10^{-6}/°C.

The changes in pore volume of a 22.3 percent porosity Berea
sandstone to changes in stress at two different temperatures is reported by Von Gonten and Choudhary (1969). At an effective stress
of 6.9 MPa and a temperature of 205°C the pore volume shows a decrease of 1.7 percent; at an effective stress of 13.8 MPa and a
temperature of 205°C the pore volume decreases by 3 percent. This
indicates that at higher effective stresses a larger decrease in
pore volume occurs. It also shows that the pore volume does decrease
with increasing temperature. The major problem in comparing Von
Gonten and Choudhary's results with the current work is that their
experiments were conducted at constant temperature levels and constant pore pressures while varying the confining stress; the experiments reported here maintain constant confining and pore pressures
while varying the temperature.

EQUIPMENT AND PROCEDURE

The experimental equipment was designed to provide a high
pressure and high temperature environment for consolidated sandstone

cores. Components of the apparatus are provided to monitor and measure axial thermal expansion and pore volume contraction. A description of the entire system and preparation procedure is described in detail by Ashqar (1979) and Greenwald (1980).

In preparation for each experiment a new 5.1 cm long by 5.1 cm diameter cylindrical core is cut, dried, and its pore and bulk volumes are determined. At no time during the preparation procedure is the core ever heated to more than 65°C. A 350 ohm strain gage is mounted onto the core's side. After all wires are well insulated, stainless steel caps are placed on each end of the sample and a shrink-fit teflon tube jackets the assembly. A high vacuum is then pulled on the core for 12 hours, after which deaerated distilled water is admitted to the sample. Complete saturation of the core is insured by measuring both the water admitted to the core and the core's weight differential before and after saturation; full saturation is of utmost importance. After these steps are completed, the strain gage leads are secured to wires which pass from a strain gage indicator to the interior of a high pressure cell.

The test equipment is shown schematically in Figure 2. It consists primarily of the high pressure test cell, two pressurizing systems, a heating system, and strain monitoring equipment. The test cell, rated to 69 MPa, contains the jacketed core sample. The core is hydrostatically stressed with silicone oil employed as the pressurizing medium. A back pressure regulator is used at the cell outlet to provide constant confining stress during heating.

The high pressure cell is surrounded by four resistance heaters coupled to a temperature controller. These heaters fit the contour of the cell and are surrounded by thermal insulation. It is necessary to maintain the heating rate below 2°C/minute as this is believed to be the threshold that causes thermal cracking of rock grains (Richter and Simmons (1974)). A heating rate of 1.5°C/minute was used in all experiments in this work.

The pore pressure system is used to flow distilled water through the core. This is done before each experiment with a back pressure of 1.3 MPa; several pore volumes of water are flowed through the core to remove trace quantities of air possibly remaining in it. When this is completed, valves A and B on Figure 2 are closed and an experiment can commence.

The first step in an experiment is to use the air-actuated pump to raise the confining stress to its desired level. Pore pressure is increased by utilizing the high pressure hand pump. This positive displacement pump is equiped with a vernier which indicates volume changes to within $\pm$ .0005 cc. With the pressures set, the heating begins, and data can be obtained. The hand pump must be continually

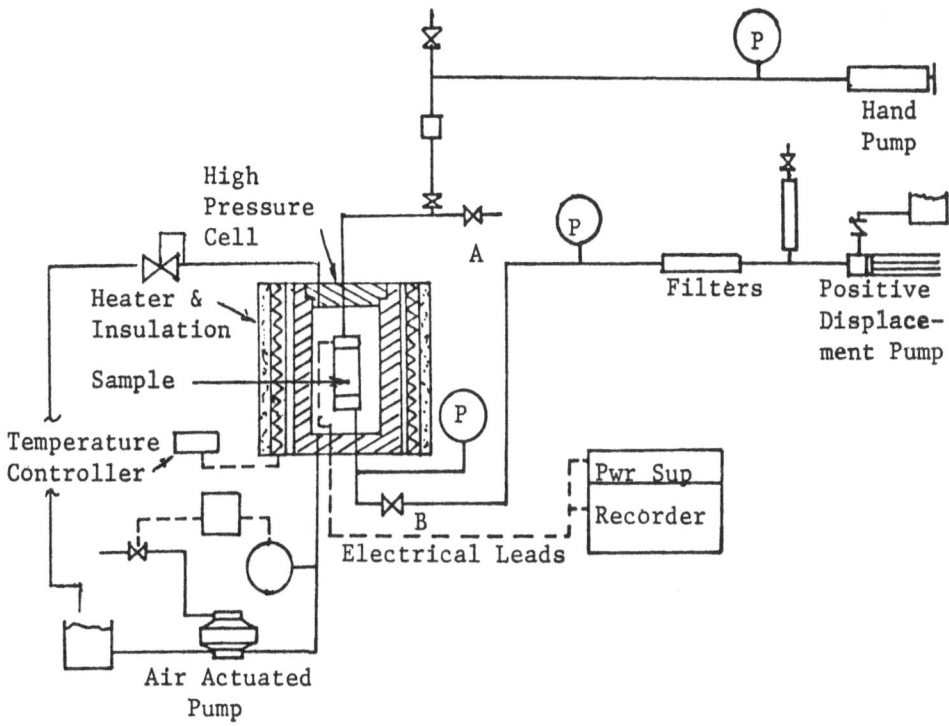

Figure 2.   Equipment schematic.

adjusted to maintain a constant pore pressure; confining stress is automatically controlled by the back pressure regulator. At approximately 10°C intervals the hand pump volume and strain gage readings are taken.

    To obtain true values for the strain, it is necessary to correct for the variation of the gage factor as a function of temperature. The maximum variation is two percent in the temperature range of the present tests. This is a straight-forward procedure and yields the axial bulk thermal expansion. The hand pump volume readings require two corrections. The first of these is an equipment calibration due to thermal expansion of the caps, tubing and water contained within them. A second correction is made to account for the thermal expansion of the water contained within the sample core space. These two corrections may be rather major in some temperature ranges amounting to as much as one-third of the apparent volume change. When these corrections are accounted for, the thermal contraction of the pore volume is found.

RESULTS

Figure 3 shows the axial expansion of the three sandstones
measured perpendicular to their bedding planes. The effective stress
for all samples is 13.8 MPa, as obtained by subtracting the pore
pressure of 6.9 MPa from the confining stress of 20.7 MPa. It is
noteworthy that the Boise sample shows nearly linear behavior. This
is to be anticipated, as Boise sandstone shows the most nearly linear
elastic behavior of the three samples (Wilhelmi and Somerton (1967)).
The Bandera sandstone shows the closest behavior to that observed by
Somerton and Selim (1961). Its axial expansion is .21 percent in
this work versus .22 percent in the previous study. This may be because it has the lowest porosity and is thus least affected by the
inclusion of pore fluids. The dissimilarity of the Boise and the
Bandera and Berea samples at low temperatures can be interpreted as
possible evidence of micro-crack closure occurring. The highly non-
linear stress-strain behavior of these latter samples as observed by
Wilhelmi and Somerton (1967) is evidence of this. Berea 1 is the
only sample shown that exhibits a decreasing thermal expansion at
higher temperatures. This trend was evident in two other Berea
samples that were tested and is not presently understood.

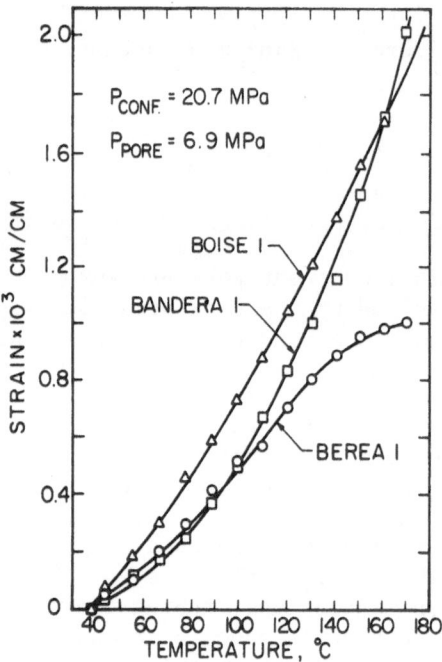

Figure 3. Axial strain versus temperature.

The cumulative percentage of pore volume contraction is shown in Figure 4. Ranking of the three sandstones shows that Bandera has the greatest pore volume contraction followed by Berea and Boise. This is also the order of increasing porosity. Thus, it appears that a less porous sandstone when heated will lose a greater percentage of its original porosity than will a more porous sample. This is expected and can be explained conceptually by envisioning a single constant-pressure pore which exists as a sphere in a non-porous spherical shell. If the shell is heated under stress and expands, it will cause the pore volume to decrease by a certain amount. By replacing the original pore with a smaller spherical pore in the same shell, it is obvious that upon heating, a greater fraction of the initial porosity is lost. The similarity in shape of all three curves is to be expected due to their similar mineralogical compositions.

Figure 5 combines the results of both pore volume and axial thermal expansion measurements to calculate decrease in the original porosity. This requires that one assume all three sandstones are isotropic and thus, that the volumetric strain is equal to three times the axial strain. By assuming that the data of Somerton and Selim is representative of the thermal anisotropy, it is found that an isotropy assumption introduces a maximum probable error of only 2 percent. Figure 5 exhibits the same behavior and trends as Figure 4, indicating that pore volume contraction is of primary importance compared to bulk volume expansion when determining porosity variations.

The result of effective stress variations on porosity reduction is shown in Figure 6. Results are given for two samples of Boise sandstone which were cored from the same block and in the same direction. The pore pressure is 6.9 MPa for both samples, while the confining stress for Boise 1 is 20.7 MPa and for Boise 2 is 13.8 MPa. The curves indicate that the higher effective stress causes a greater reduction in porosity. This is best explained by noting that the higher effective stress implies a lower pore pressure in relation to the stress existing throughout the solid. Therefore, there is less resistance to grain expansion into the pore space than to displacing adjacent grains.

All of the results given in this paper are based on one sample each of Berea and Bandera sandstones and two samples of Boise. Reproducibility runs for the samples shows very good qualitative agreement but with the numerical results showing some differences. The source of these differences is not presently known and work is continuing in order to evaluate them.

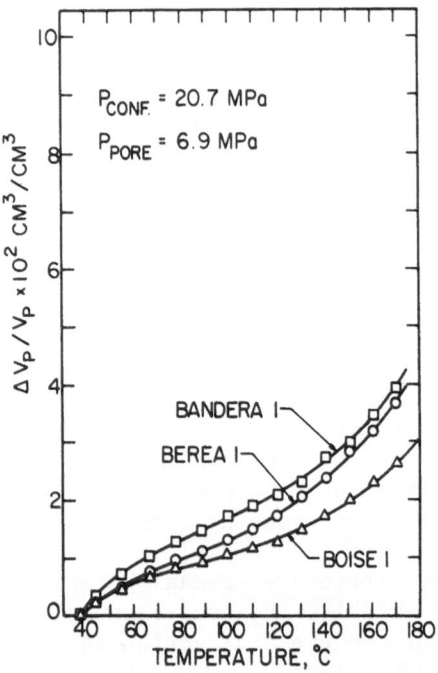

Figure 4. Fractional pore volume change versus temperature.

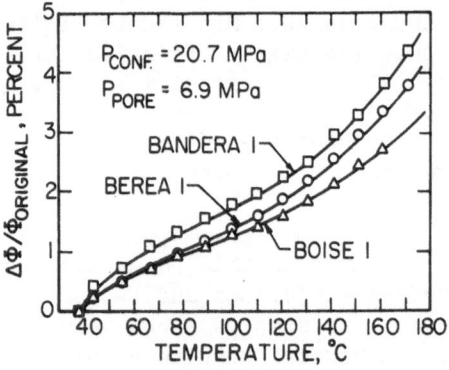

Figure 5. Fractional porosity change versus temperature.

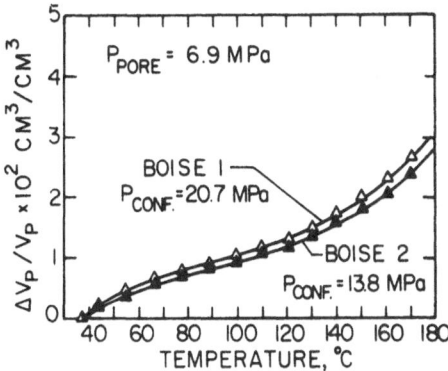

Figure 6.   Effect of stress on fractional pore volume change.

CONCLUSIONS

The decrease in pore volumes with increased temperature for
the sandstones used in this work are in general agreement with results of Von Gonten and Choudhary (1969). Two major trends are
noted: (1) the cumulative percentage decrease in pore volume increases at any temperature with increasing effective stress, and
(2) the cumulative percentage decrease in pore volume at any stress
increases with increasing temperature.

The linear thermal expansions of Bandera and Boise sandstones
agree in general with the earlier results of Somerton and Selim
(1961) for dry sandstones under zero stress. Results for Berea
sandstone did not agree and are the subject of further investigations.

Calculated changes in porosity with temperature based on measured bulk volume expansion and pore volume contraction agreed
favorably with those calculated by Somerton and Mathur (1976). The
fractional change in porosity is very similar to the results for
fractional pore volume change; this is because the inclusion of
bulk volume thermal expansion values in porosity determinations is
a second-order correction.

The information obtained in this work has numerous applications.
One use is in the study of the effect of temperature on the permeability of sandstones. It has been found by Wong (1979) that increasing the temperature of a porous rock causes its permeability to
decrease. The decrease in pore volume with temperature as observed

in this study may aid in the explanation of this phenomenon.

A second application of the results of this work is the relating of surface-measured porosities of rock samples to their in-situ values. By knowing the temperature of the reservoir that a core sample is taken from, the thermal contraction of its pore space can be quantified.

Thirdly, it is possible that the driving force which pushes oil to the surface is partly due to the compressibility of the formation itself. This is usually the case for reservoirs which are undersaturated. In such situations it is possible that heating the formation will produce an additional drive mechanism due to the reduction in porosity of the reservoir rock. The present work provides a first step in determining the magnitude of this factor.

The larger decrease in pore volume which arises as the effective stress is increased is believed to occur for all three sandstones. More complete work is needed, however, at three or four stress levels to accurately quantify this effect. Experiments are continuing in this direction.

ACKNOWLEDGMENTS

This work was made possible in part by a grant from Getty Oil Company. Support for the work was also provided by the Department of Energy on Lawrence Berkeley Laboratory Contract Number W-7405-ENG-48.

REFERENCES

Ashqar, P. I., 1979, Thermal Expansion of Porous Rocks Under Stress, Master of Science Project Report, University of California, Berkeley, December 1979.

Greenwald, R. F., 1980, Ph.D. dissertation to be published, University of California, Berkeley, 1980.

Harvey, R. D., 1967, Thermal Expansion of Certain Illinois Limestones and Dolomites, Ill. St. Geol. Surv., Circular 415, p. 1-33.

Mann, R. L. and Fatt, I., 1960, Effects of Pore Fluids on the Elastic Properties of Sandstone, Geophysics, Vol. 25, No. 2, April 1960, p. 433-444.

Ritcher, D. and Simmons, G., 1974, Thermal Expansion Behaviour of Igneous Rocks, Int. Jour. Rock Mech. Min. Sci. and Geomech., Vol. II, p. 403-416.

Skinner, B. J., 1966, Thermal Expansion in "Handbook of Physical of Constants," Geol. Soc. of Amer. Memoir No. 97, p. 75-96.

Somerton, W. H. and Mathur, A. K., 1976, Effects of Temperature on Fluid Flow and Storage Capacity of Porous Rocks, Proceedings 17th U.S. Symposium on Rock Mechanics, Snowbird, Utah, August 25-27, 1976.

Somerton, W. H. and Selim, M. S., 1961, Additional Thermal Data for Porous Rocks -- Thermal Expansion and Heat of Reaction, Soc. Pet. Engrs. Jour., Vol. 4, December 1961, p. 249-253.

Von Gonten, W. D. and Choudhary, B. K., 1969, The Effect of Pressure and Temperature on Pore Volume Compressibility, Soc. Pet. Engrs. Preprint 2526, September, 1969.

Wilhelmi, B. and Somerton, W. H., 1967, Simultaneous Measurement of Pore and Elastic Properties of Rocks Under Triaxial Stress Conditions, Soc. Pet. Engrs. Jour., Vol 7, No. 3, September, 1967, p. 283-294.

Wong, L., 1979, The Effect of Temperature on Permeability, Master of Engineering Project Report, University of California, Berkeley, December, 1979.

THERMAL EXPANSION AND MAGNETIC PROPERTIES OF Fe-Pd INVAR ALLOYS

CONTAINING CARBON

K. Fukamichi*, M. Kikuchi[†], and T. Nakayama[†]

The Research Institute for Iron, Steel and Other Metals
Tohoku University, Sendai, 980, Japan

ABSTRACT

In order to study the thermal expansion characteristics and the magnetic properties of Fe-Pd Invar alloys, the γ-phase was stabilized by addition of carbon. Since the carbon addition makes cold working easy, the effect of cold working on the thermal expansion characteristics was also investigated. The Curie temperature and the saturation magnetization were almost independent of the carbon content. The thermal expansion anomaly below the Curie temperature became more remarkable by addition of carbon, and it was significantly pronounced by cold working, as indicated by a negative thermal expansion in a wide temperature range. It was found from the thermal expansion hysteresis curves that the γ-phase of the alloys containing carbon is more stable than that of $Fe_{70}Pd_{30}$ binary alloy. The linear magnetostriction at room temperature showed a large positive value by quenching or addition of carbon. The ΔE effect became small by addition of carbon in the quenched state.

INTRODUCTION

Up to the present time, many kinds of Invar alloys have been

*Present temporary address: IBM Thomas J. Watson Research Center,
 Yorktown Heights, NY 10598

[†]The Research Institute of Electric and Magnetic Alloys, Sendai,
 982, Japan

developed[1]. These Invar alloys have a relatively low Curie temper-
ature and their composition is very close to the γ(f.c.c) $\rightleftharpoons$ α(b.c.c)
phase transformation. The studies of the Invar effects of Fe-Pd
alloys have been not so active[2,3] because the phase diagram is
somewhat complex[4,5]. That is to say, at high temperature range,
this alloy system has a γ-phase at Fe-rich region, but the γ-phase
decomposes into α-Fe and FePd superstructure[4,5]. Then, we can not
obtain Invar effects in the annealed state. The rapidly cooled and
cold worked specimens show Invar characteristics, but they are not
stable[2]. The magnetic properties such as the concentration dependence of the magnetic moment reported previously differ from each
other due to the unstable state of the γ-phase[3,6].

     The γ-phase in Fe-rich region, as is well known, is stabilized
by addition of carbon[7,8]. The effect of carbon on the magnetic
properties of Fe-Ni Invar alloys has been investigated by many
workers, and many important results have been obtained[9~12]. Then,
it is expected that the addition of carbon is also useful for the
study of Invar characteristics of Fe-Pd alloys. In the present
paper, firstly, the thermal expansion characteristics and the
magnetic properties of Fe-Pd alloys with γ-phase stabilized by
addition of carbon are reported. It has been known that the thermal
expansion anomaly of Fe-Ni Invar alloys is enhanced by cold working
[13~16]. The cold working of Fe-Pd binary alloys is difficult and so
Jessen has carried out the cold working up to only 1%[2]. However,
the cold working of Fe-Pd alloys containing carbon is relatively
easy. Next we have investigated the effect of cold working on the
thermal expansion characteristics. Finally, the effects of quenching and addition of carbon on the elastic properties and the lienar
magnetostriction are reported. In this paper, the alloy compositions are given as nominal atomic percent.

EXPERIMENTAL

     The several kinds of Fe-Pd-C alloys were prepared by arc-
melting in an argon atmosphere. These alloys were remelted in an
argon atmosphere and sucked into a silica tube. In order to
homogenize, the annealing was done at 1000°C for 15~24 hr. The
specimens were quenched into an iced brine. The cold working was
carried out at room temperature after quenching. The thermal
expansion was measured with the dilatometer construction (Rigaku
Thermoflex 8561 Al). The magnetization was measured by a Farady
method, and the saturation magnetization and the Curie temperature
were determined. Young's modulus was measured by an electrostatic
driving method[17]. The linear magnetostriction was measured with
the apparatus devised by Honda et al[18].

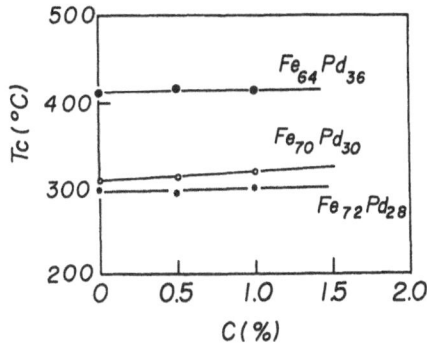

Fig. 1.   The Curie temperature as a function of the carbon
concentration for Fe-Pd alloys.

RESULTS AND DISCUSSION

     Figure 1 shows the Curie temperature as a function of the
carbon concentration. In the case of Fe-Ni Invar alloys, the Curie
temperature and the saturation magnetization increases linearly
with carbon content[10,11], accompanied by an increase in the lattice
constant[11]. In addition, the hydrostatic pressure dependence of the
Curie temperature is very remarkable. For instance, the value of
$\partial T_C/\partial P$ for $Fe_{64}Ni_{36}$ Invar alloy is about -3.4 K/kbar[19,20]. These
results suggest that the magnetic properties such as the Curie
temperature depend sensitively on the lattice constant. In the
case of Invar alloys, the lattice constant does not vary linearly
with the content, deviating strongly from Vegard's law[22], and the
lattice constants at room temperature of Fe-Pd Invar alloys are
originally larger than by about several percent that of Fe-Ni Invar
alloys[23]. Then, the change in the Curie temperature is not so
remarkable as shown in the figure. This fact is suggestive that the
effect of carbon on the lattice expansion for Fe-Pd alloys is less
than that for Fe-Ni alloys. We have also measured the saturation
magnetization, but the effect of carbon on it is not detectable.
This behavior would be attributed to the same reason mentioned above.

     The effect of quenching on the thermal expansion characteris-
tics is shown in Fig. 2, together with the curves of $Fe_{69}Pd_{31}$ and
$Fe_{71}Pd_{29}$ alloys cooled from 900 °C at the rate of 50°C/min[2]. The
curve of the quenched $Fe_{70}Pd_{30}$ alloy shows more remarkable anomaly
than those of the cooled specimens. Similar behavior has been
observed in Fe-Ni Invar alloys[21]. These results imply that the
perfect γ-phase of Fe-Pd alloys is obtainable by rapid quenching.

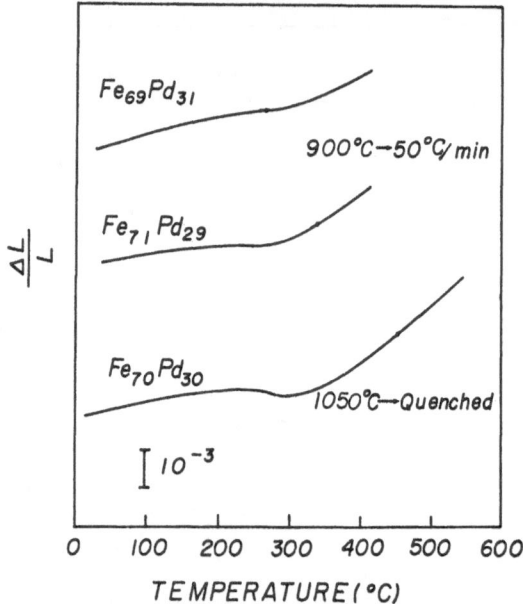

Fig. 2.   The effect of heat treatment on the thermal
expansion characteristics of Fe-Pd alloys.

        Figure 3 shows the thermal expansion curves of Fe-Pd alloy
system. The ratio of R = Pd/(Fe+Pd) was fixed at 0.3. With
increasing carbon content, the thermal expansion anomaly becomes
more remarkable, and in the case of the alloy with 1.5% C, a
negative thermal expansion is observed around 250°C. It has been
reported that the thermal expansion anomaly in Fe-Ni Invar alloys
is pronounced by quenching[21], while the addition of carbon increases
thermal expansion coefficient[24]. Therefore, the effect of carbon
on the thermal expansion anomaly for Fe-Pd alloys is in contrast to
that of Fe-Ni Invar alloys. As is well known, the Invar characteristics of Fe-base alloys are peculiar behaviors in the γ-phase, so
the remarkable thermal expansion anomaly of Fe-Pd alloys would be
caused by the stabilization of γ-phase by addition of carbon. That
is to say, as mentioned at the beginning, the γ-phase of Fe-Pd
alloys is unstable, then the carbon addition leads a stabilization of
the γ-phase which brings about a thermal expansion anomaly rather
than acts a impurity which results in a increase in the thermal
expansion coefficient. The cooling curves do not show the thermal
expansion anomaly, it means that the γ-phase decomposes into mainly
α-Fe and FePd by heating above about 600°C.

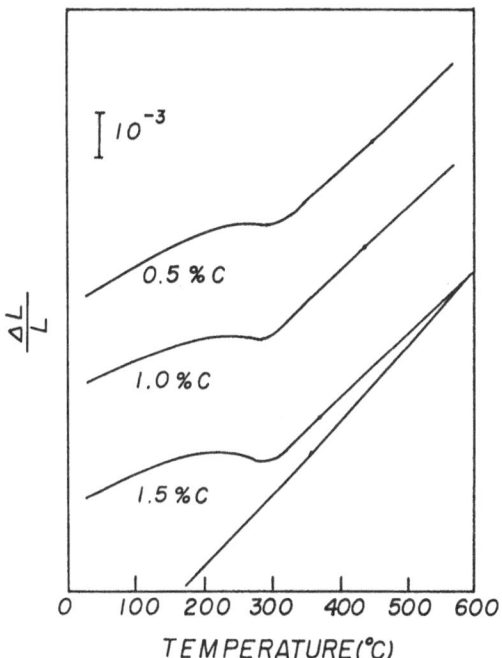

Fig. 3.   The thermal expansion curves of Fe-Pd alloys
          (R=0.3) containing carbon.

        The cold working of Fe-base γ-phase alloys becomes more dif-
ficult with increasing Fe content. The cold working of $Fe_{69}Pd_{31}$
alloy has been carried out up to only 1%[2]. However, by addition
of carbon, the cold working is easily made. In the present study,
the alloys (R=0.3) containing carbon have been cold-rolled up to
about 20%. The effect of cold working on the thermal expansion
characteristics is shown in Fig. 4. The alloy with no carbon does
not show a large negative thermal expansion by quenching, but the
alloy with 1% C shows a large anomaly below the Curie temperature,
and this anomaly is drastically enhanced by cold working, indicating
a negative thermal expansion in a wide temperature range. However,
as shown in the figure, at higher cold working level, the thermal
expansion coefficient shows a large positive value, although a
drastic shrink takes place around 400°C. Similar curves of Fe-Ni
swaged alloys have been obtained by Tino and Kobayashi[15]. They
have pointed out that the temperature region of the α ⇄ γ transformation of Fe-Ni Invar alloys shifts considerablly to the higher
temperature side[15]. Especially, as seen in the phase diagram[3], the
transformation temperature is very close to room temperature at 30%
Pd, and so the strain-induced martensitic transformation tends to

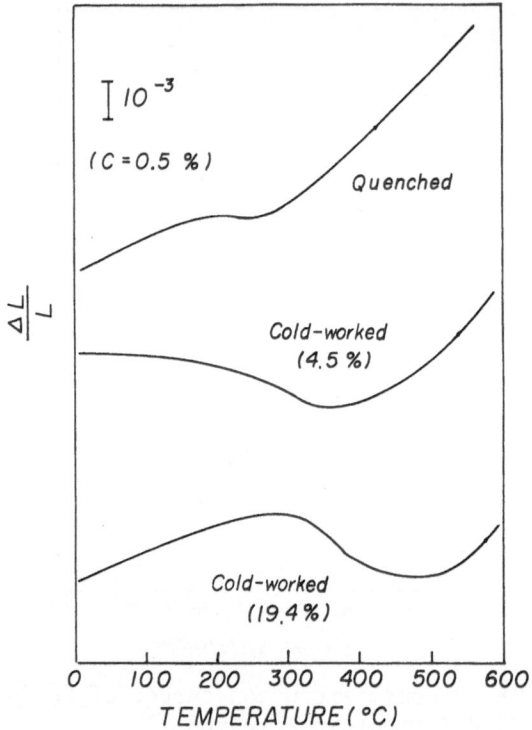

Fig. 4. The thermal expansion curves of cold-worked Fe-Pd
alloys (R=0.3) containing 0.5% carbon.

occur by cold working. Therefore, the drastic change in the thermal
expansion anomaly for the specimen cold-worked harder would be
correlated to such a phase transformation.

Jessen has cold-worked $Fe_{69}Pd_{31}$ alloy and measured the heating
and cooling curves of thermal expansion , and he has pointed out
that this alloy is very unstable, and the thermal expansion curves
show a thermal hysteresis by heating even up to 180°C.

From the results of Figs. 3~4, it is concluded that the thermal
expansion anomaly below the Curie temperature is enhanced by
addition of carbon and cold working. In the present paper, the

value of R was fixed at 0.3. We have also found similar behaviors
in other compositions. The enhancement of thermal expansion by
cold working has been recently discussed by several investigators.
Tino and Kobayashi have explained it by using the shear mechanism
of $\gamma \rightleftharpoons \alpha$ martensitic transformation[15]. Kagawa and Chikazumi have
discussed this phenomenon by taking into consideration the
anisotropic Fe-Fe pair distributions induced by cold working[16]. As
is well known the magnetic properties of alloys with Invar region
are sensitively affected by environment effects[25]. Kanamori has
discussed the Invar characteristics by using a coherent potential
approximation method, and pointed out that Fe-Fe pairs are responsible for the Invar effect[26]. His model can also explain such an
enhancement of the thermal expansion anomaly mentioned above because
the cold working causes the change in the number of Fe-Fe pairs. In
order to discuss in more detail the enhancement of the thermal
expansion anomaly of Fe-Pd alloys containing carbon, the experiments
such as the magnetic anisotropy and the Mössbauer effect are
necessary, in addition, the martensitic transformation temperature
should be determined exactly in the Fe-Pd ternary alloys containing
carbon.

Other magnetic properties of Fe-Pd alloys are changed by
quenching or addition of carbon. Table-1 shows the linear magneto-
striction $\lambda_{\prime\prime}$ and the ΔE effect at room temperature for the alloys
with R = 0.3. It has been already found that the value of $\lambda_{\prime\prime}$
becomes large by quenching[27]. The value of $\lambda_{\prime\prime}$ of the alloy with 0.5
% C is larger than that of the slow cooled specimen by a factor of
about 2. From the practical point of view, it is desirable to
obtain a large magnetostriction in a low magnetic field H. The
value of $d\lambda_{\prime\prime}/dH$ below 100 Oe for Fe-Pd alloys containing carbon is
larger than that of $Fe_{50}Co_{50}$ conventional magnetostriction alloy[28].
On the other hand, the ΔE effect of the alloy containing carbon is
smaller than that of the Fe-Pd binary alloy[28]. As is well known,
the relationship between the saturation magnetostriction and the ΔE
effect is given by the following equation[30].

Table-1.   The linear magnetostriction and the ΔE effect of Fe-Pd
binary alloy and its alloy containing carbon.

	$Fe_{70}Pd_{30}$	R (=0.3)+0.5%C
$\lambda_{\prime\prime}(\times 10^{-6})$	27 (slow cooled)	89 (slow cooled)
	70 (quenched)	62 (quenched)
$\Delta E(\%)$	6.0 (quenched)	4.3 (quenched)

$$9\lambda_s{}^2 E_s \mu_0 = 20\pi M_s{}^2 (\frac{\Delta E}{E_D})$$

where E_s and E_D are Young's moduli at saturated and zero magnetic field, respectively, μ_0 the initial permeability and M_s the saturation magnetization. The ΔE effect is defined by $\Delta E = (E_S - E_D)/E_D$. In the case of Fe-Ni alloys, the magnitude of the linear magnetostriction is roughly proportional to that of the ΔE effect[31]. In the present alloy system, such a relation is also observed.

SUMMARY

The γ-phase of Fe-Pd binary Invar alloys is unstable, therefore in the present study, the addition of carbon has been carried out in order to stabilize the γ-phase. The thermal expansion characteristics and the magnetic properties for quenched and cold-worked specimens have been investigated. The main obtained results are as follows;

(a) The saturation magnetization and the Curie temperature are hardly affected by addition of carbon.

(b) The thermal expansion characteristics is sensitive to the quenching or cooling condition.

(c) By addition of carbon, the cold working is easily done.

(d) The thermal expansion anomaly below the Curie temperature becomes more remarkable by addition of carbon, and its anomaly is significantly enhanced by cold working, but further cold working brings about a positive large thermal expansion coefficient around room temperature and a drastic shrinkage around 400°C.

(e) The linear magnetostriction and the ΔE effect of Fe-Pd alloys are increased by quenching or addition of carbon.

ACKNOWLEDGMENTS

The authors are grateful to Dr. H. Masumoto of the Research Institute of Electric and Magnetic Alloys for his continuing encouragements and supports during the present study. The arc-melting was done at the laboratory of Prof. T. Wakiyama of the Department of Electrical Engineering, Tohoku University.

REFERENCES

1. *PHYSICS AND APPLICATIONS OF INVAR ALLOYS*, ed. H. Saito, Maruzen Co. Ltd. Tokyo (1978).
2. K. Jessen, Ann. Phys. (Leipzig) 9, 313(1962).
3. A. Kussmann and K. Jessen, J. Phys. Soc. Japan, Suppl. 17-B1, 136(1962).

4. E. Raub, H. Beeskow and O. Loebich Jr., Z. Metallk., 54, 549 (1963).
5. A. Kussmann and K. Jessen, Z. Metallk., 54, 504(1963).
6. H. Fujimori and H. Saito, J. Phys. Soc. Japan, 20, 293(1965).
7. B. S. Lement, B. I. Averbach and M. Cohen, Trans. ASM, 43, 1072(1951).
8. G. F. Bolling and R. H. Richman, Phil. Mag., 19, 247(1969).
9. I. Y. Georgieva and O. P. Maksimova, Phys. Met. Metallogr., 24(3), 200(1967).
10. G. F. Bolling, A. Arrott and R. H. Richman, Phys. Stat. Sol., 26, 743(1968).
11. E. Adler and C. Radeloff, Z. Angew. Phys., 26, 105(1969).
12. R. Caudron, J. J. Meunier and P. Costa, J. Phys. F, 4, 1791 (1974).
13. R. Schulze, Z. Angew. Phys., 7, 57(1955).
14. G. Pupke, Z. Phys. Chem., 207, 91(1957).
15. Y. Tino and M. Kobayashi, J. Phys. Soc. Japan, 41, 59(1976).
16. H. Kagawa and S. Chikazumi, J. Phys. Soc. Japan, 43, 1097 (1977).
17. Y. Shirakawa and I. Oguma, Sci. Rep. RITU, A-18S, 523(1966).
18. K. Honda, H. Masumoto, Y. Shirakawa and T. Kobayashi, J. Japan Inst. Metals, 12, 1(1948).
19. L. Patrick, Phys. Rev., 93, 384(1954).
20. J. M. Leger, C. Loriers-Susse and B. Vodar, Phys. Rev., B6, 4250(1972).
21. Y. Tino and M. Kobayashi, J. Phys. Soc. Japan, 45, 1226(1978).
22. M. Shiga, Solid State Commun., 10, 1233(1972).
23. W. B Pearson, *A Handbook of Lattice Spacings and Structures of Metals and Alloys*, Pergamon Press, New York, London (1967) 634.
24. Ch. Ed. Guillaume, Compt. Rend, 170, 1433(1920).
25. J. Kanamori, J. Phys., 35, C4-131(1974).
26. J. Kanamori, *PHYSICS AND APPLICATIONS OF INVAR ALLOYS*, ed. H. Saito, Maruzen Co. Ltd, Tokyo (1978) 221.
27. K. Fukamichi, J. Appl. Phys., 50, 6562(1979).
28. Y. Mashiyama, Sci. Re. RITU, 21. 394(1932).
29. T. Nakayama, M. Kikuchi and K. Fukamichi, J. Phys. F, 10, 715 (1980).
30. M. Kersten, Z. Phys., 85, 708(1933).
31. R. M. Bozorth, *Ferromagnetism*, D. Van Nostrand, New York (1951) 684.

ANALYSIS OF THERMALLY GENERATED MICROSTRESSES IN POLYCRYSTALLINE
Be DUE TO THE PRESENCE OF BeO INCLUSIONS

T. A. Hahn

Center for Materials Science
National Bureau of Standards
Washington, D.C. 20234

Engineering Materials Group
Chemical Engineering Department
University of Maryland
College Park, Maryland 20742

R. W. Armstrong
Engineering Materials Group, and
Mechanical Engineering Department
University of Maryland
College Park, Maryland 20742

ABSTRACT

In polycrystalline hexagonal beryllium (Be), the microstresses
due to the relatively small thermal expansion anisotropy were
previously estimated, on an ideal elastic basis, to be comparable
to those stresses measured for the general yield and fracture
strengths of bulk Be material. Commercial beryllium materials
normally contain inclusions of (hexagonal) beryllium oxide (BeO),
and the mismatch of expansivity between these materials is now
shown to be capable of producing even larger elastic microstresses
which should produce additional localized yielded and fracture
zones within the material. Despite the fact that the BeO inclusions are in compression, both compressive and tensile stresses
are generated in the surrounding shell of Be material. The
plastic zone size is estimated to be as much as ten times larger
than an inclusion diameter so that an important parameter affecting the nature of plastic flow and cracking around an inclusion should be the polycrystal grain size. These several consid-
erations are described on the basis of plasticity and dislocation
models which are proposed for the material behavior.

INTRODUCTION

The thermal microstresses generated between individual grains within pure polycrystalline Be material have been calculated on an ideal elastic basis by Armstrong and Borch[1]. It was shown for a polycrystalline aggregate having Reuss' average compliances and an average thermal expansivity that a single misfit grain within the average matrix would generate thermal microstresses comparable to the yield stress and even the fracture stress of bulk Be material. Pope and Stevens[2], in comparing their own measurements of the yield behavior of shock loaded Be single crystals to measurements on polycrystalline Be, found the seemingly anomalous result that the polycrystalline Be had a lower yield strength. They suggested that this yielding occurred in polycrystalline material at lower external stress values because of the presence of internal thermal micro-stresses of the type described by Armstrong and Borch.

Burns, Gurland, and Richman[3] have observed cracking at in-clusions in bending experiments on polycrystalline Be material. Commercial Be generally contains on the order of 5% BeO. Be and BeO were reported by Armstrong and Borch to have rather different compliances and thermal expansivities so it seemed of interest to consider whether thermal microstresses might also be important in determining the mechanical effects of BeO inclusions in Be. The following calculations show that relatively large stresses and consequent deformation and possible cracking effects should result from the mismatched properties of these phases.

THE ELASTIC THERMAL MICROSTRESSES

Table 1 shows the melting temperatures[1], compliances[4] at 300 K, and thermal expansivities[5] at 300 K for Be and BeO crystals. Table 2 contains the Reuss' average compliances, Poisson's ratios and average thermal expansivities at 300 K which were calculated from the data in Table 1.

Table 1. Melting Temperatures, Compliances, and Thermal
         Expansivities of Be and BeO at 300 K.

	T_m	S_{11}	S_{33}	S_{12}	S_{13}	S_{44}	α_a	α_c
	(K)		$(10^{-12} \ Pa^{-1})$				$(10^{-6} \ K^{-1})$	
Be	1556	3.46	2.98	−0.31	−0.13	6.15	12.3	8.9
BeO	2803	2.40	2.15	−0.60	−0.32	6.77	7.0	5.9

Table 2. Reuss' Average Compliances, Poisson's Ratios, and
          Average Thermal Expansivities of Be and BeO at 300K.

	$\overline{S}_{11}$	$\overline{S}_{12}$	$\overline{S}_{44}$	$\overline{\nu}$	$\overline{\alpha}$
	$(10^{-12}\ Pa^{-1})$				$(10^{-6}\ K^{-1})$
Be	3.23	−0.15	6.76	0.05	11.2
BeO	2.52	−0.52	6.09	0.21	6.6

The individual Be and BeO materials are not especially
anisotropic. The anisotropy factor, $A = 2(S_{11} - S_{12})/S_{44}$, has a
value of 1.2 for Be and 1.6 for BeO. However, the Tables show
there is a large mismatch in the thermal expansivities between
the materials which indicates a potential driving force for
generating exceptional internal stresses when cooling these phases
in mutual contact.

The elastic stresses in anisotropic materials can be calculated from the generalized tensor form of Hooke's law as given
in Equation 1.

$$\varepsilon_{ij} = S_{ijkl}\ \sigma_{kl} \tag{1}$$

The strain components, ε_{ij}, are determined by the differences in
expansions of the Be and BeO phases achieved by cooling from the
temperature at which the sample is either manufactured or equi-
librated by heat treatment. Various cases can be considered, such
as: a single crystal BeO inclusion in a single crystal Be matrix;
a polycrystalline BeO inclusion in a single crystal Be matrix;
or, a single crystal BeO inclusion in a polycrystalline Be matrix.

Equation 2 specifies the thermal strain in the 1 direction,
i.e. [2$\overline{1}\overline{1}$0] for a single crystal BeO inclusion in a polycrystalline Be matrix.

$$\left(\overset{Be}{\underset{\alpha}{\ }} - \overset{BeO}{\underset{\alpha_1}{\ }}\right)\Delta T = \left(\overline{S}_{11}^{\,Be} + S_{11}^{\,BeO}\right)\sigma_{11} + \left(\overline{S}_{12}^{\,Be} + S_{12}^{\,BeO}\right)\sigma_{22} + \left(\overline{S}_{13}^{\,Be} + S_{13}^{\,BeO}\right)\sigma_{33}$$

$$\tag{2}$$

Two similar equations are obtained for the strains in the 2 and 3 directions, i.e. $[01\bar{1}0]$ and $[0001]$, respectively. These three equations can be rearranged to give the two stresses specified in Equations 3 and 4.

$$\sigma_{11} = \sigma_{22} = \frac{\left(\frac{Be}{\alpha} - \alpha_1\right)\Delta T\left(\bar{S}_{33}^{Be} + S_{33}^{BeO}\right) - \left(\frac{Be}{\alpha} - \alpha_3\right)\Delta T\left(\bar{S}_{13}^{Be} + S_{13}^{BeO}\right)}{\left(\bar{S}_{33}^{Be} + S_{33}^{BeO}\right)\left[\bar{S}_{11}^{Be} + S_{11}^{BeO} + \bar{S}_{12}^{Be} + S_{12}^{BeO}\right] - 2\left(\bar{S}_{13}^{Be} + S_{13}^{BeO}\right)^2}$$

$$(3)$$

$$\sigma_{33} = \frac{\left(\frac{Be}{\alpha} - \alpha_3\right)\Delta T\left(\bar{S}_{11}^{Be} + S_{11}^{BeO} + \bar{S}_{12}^{Be} + S_{12}^{BeO}\right) - 2\left(\frac{Be}{\alpha} - \alpha_1\right)\Delta T\left(\bar{S}_{13}^{Be} + S_{13}^{BeO}\right)}{\left(\bar{S}_{33}^{Be} + S_{33}^{BeO}\right)\left[\bar{S}_{11}^{Be} + S_{11}^{BeO} + \bar{S}_{12}^{Be} + S_{12}^{BeO}\right] - 2\left(\bar{S}_{13}^{Be} + S_{13}^{BeO}\right)^2}$$

$$(4)$$

Equations 3 and 4 are the extension to a two phase material of the previous Armstrong and Borch single phase anisotropic description. It should be noted that the stresses computed by Armstrong and Borch differ somewhat from the stress analysis sometimes quoted from the work of Likhachev[6] for an anisotropic polycrystalline aggregate, presumably because the radial equilibrium of stresses is taken into account by Likhachev. If the average values for the compliances and expansivities are employed in the latter analysis, the resultant equations agree with the equations given by Gurland[7] or Burgreen[8] for the pressure generated around an isotropic spherical inclusion in a different isotropic matrix.

Figure 1 is a micrograph of the (tensile) surface cracking observed by Burns, Gurland and Richman[3] within the polycrystal microstructure of grains for hot-pressed beryllium material in three-point bending. The axis of rotation for bending is vertical in the Figure. Of particular interest in the Figure are the individual cracks obviously centered on two inclusions at A and B. At A, the crack passes around the top side of the matrix-inclusion interface and appears to follow a Be grain boundary on either side of the inclusion. At B, a crack passes through an apparently smaller inclusion and follows a transgranular cleavage path within an individual Be grain, though, in this case the total length of the crack trace seems quite near the lower boundary of the grain. A few additional cases of apparent inclusion-associated cracks can be observed in Figure 1 but numerous other inclusions which appear

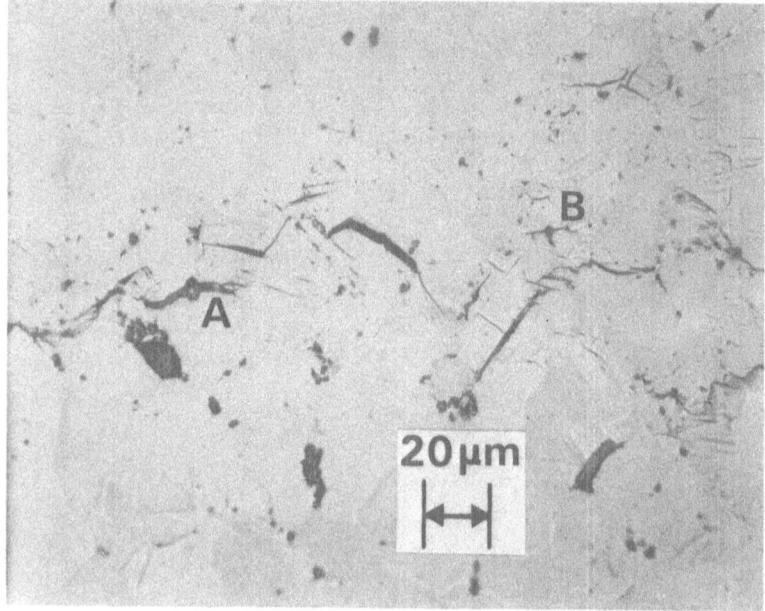

Figure 1. The tension surface of a bend specimen of hot pressed
           beryllium material, 570X. Burns, Gurland, and Richmar
           (1971) Reference 3.

free of cracking, at this magnification, are also observed. The
Figure shows clearly that certain inclusions can be associated
with the cracking process in strained material. A main consideration of the current study is to determine whether the presence
of internal microstresses due to the elastic and thermal expansion
differences of the Be and BeO materials might contribute in an
important way to such cracking, even to the extent that microcracking might be expected to occur without the addition of
external stresses.

One part of Figure 2 shows the general stresses which are
produced on a spherical shell of Be matrix surrounding a BeO
inclusion. For an average (isotropic) BeO inclusion in an average
homogeneous Be matrix, Gurland[7] has described a hydrostatic
state of stress for the mismatched inclusion as expressed by

$$\sigma_{rr}^{BeO} = \frac{(\bar{\alpha}^{Be} - \bar{\alpha}^{BeO}) \; \Delta T}{(1/2) \; (\bar{S}_{11}^{Be} - \bar{S}_{12}^{Be}) + (\bar{S}_{11}^{BeO} + 2\bar{S}_{12}^{BeO})} \qquad (5)$$

and $\sigma_{rr}^{BeO} = \sigma_{\theta\theta}^{BeO} = \sigma_{\phi\phi}^{BeO}$.

The inclusion is in a state of hydrostatic pressure since $\bar{\alpha}^{Be} >$
$\bar{\alpha}^{BeO}$, ΔT is negative, and, also, the total sum of compliance
terms in the denominator is positive.

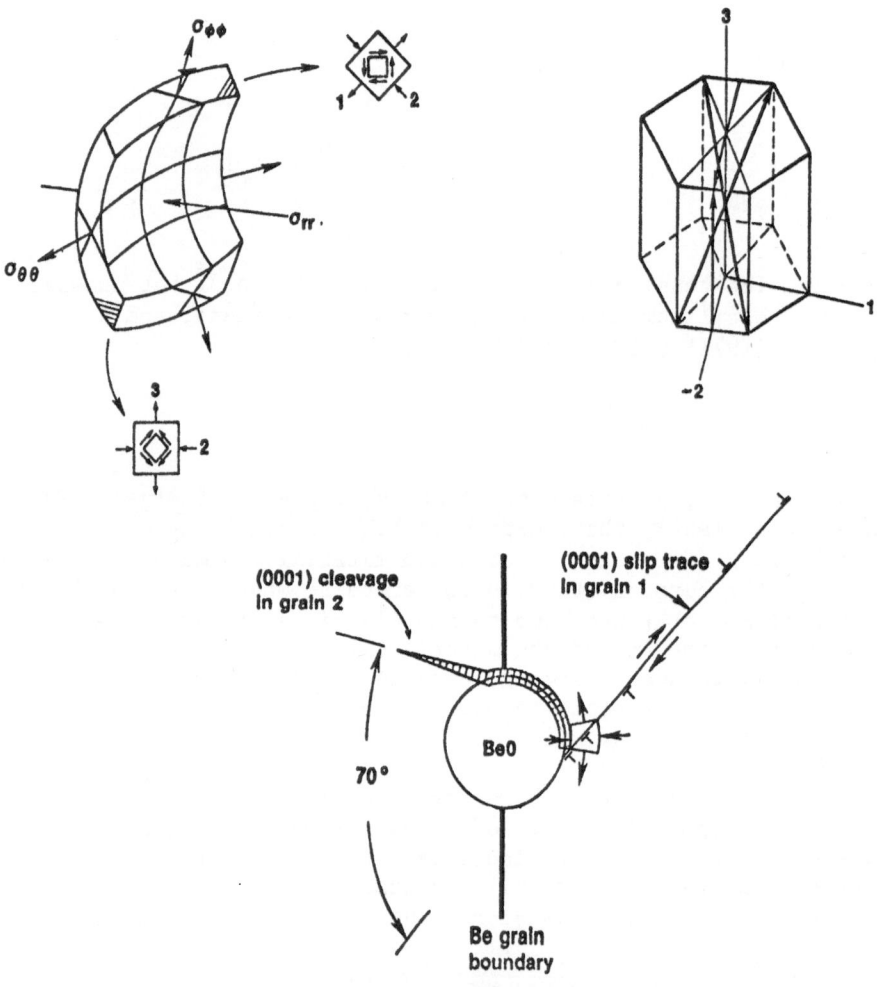

Figure 2.   Stress state in the Be matrix, crystallography of the
             Be structure, and proposed slip and cracking due to
             the thermal strains.

Gurland[7] has described the dependence of σ_{rr}, $\sigma_{\theta\theta}$ and $\sigma_{\phi\phi}$ on the distance, r, from the center of the inclusion for an elastically stressed system as

$$\sigma_{rr}^{Be} = \sigma_{rr}^{BeO} (a/r)^3$$

and   (6)

$$\sigma_{\theta\theta}^{Be} = \sigma_{\phi\phi}^{Be} = -(1/2)\ \sigma_{rr}^{BeO} (a/r)^3$$

where a is the radius of the inclusion. The radial component of stress in the matrix is compressive while the circumferential stresses, $\sigma_{\theta\theta}$ and $\sigma_{\phi\phi}$, are tensile. The difference between σ_{rr} and either $\sigma_{\theta\theta}$ or $\sigma_{\phi\phi}$ produces a shear stress as shown in the square shapes for the matching 1-2 or 2-3 planes for the r-θ or r-φ planes, respectively, in Figure 2.

Depending on the level of the shear stresses, plastic flow processes might be forced to occur within an individual grain on those slip systems potentially operable for the hexagonal cell which is also shown in Figure 2. A particular slip geometry is shown to provide the minimum number of 4 slip systems within one Be grain necessary to produce the required "thinning" deformation for a spherical shell of the matrix material. If the shear stresses are reasonably large, it might be further imagined, for example, that dislocation motion might occur on an (0001) slip system in grain 1 to sufficient extent that a crack would form at the matrix-inclusion interface in this grain, then propagate around the interface through grain 1 and grain 2, and finally switch onto a preferred (0001) cleavage plane in grain 2 according to the favorable orientation dependence of cracking at the tip of a slip band pile-up which has been previously described by Stroh[9].

Table 3 gives the temperature coefficient of the elastic stresses which should occur in the absence of plastic yielding or fracture at a matrix-inclusion interface for several types of matrix-inclusion systems. The first case is for a single crystal Be inclusion in an average polycrystalline Be matrix as described previously by Armstrong and Borch[1]. The second case is computed on the same basis for a polycrystalline BeO inclusion in a single crystal Be matrix as described by Equations 3 and 4. The difference of the principal stress coefficients in Table 3 gives a hypothetical Tresca yield stress coefficient which, when multiplied by a reasonable cooling temperature change, results in a theoretical elastic stress value near to the yield stress value measured for a variety of Be materials. For the third case in Table 3 of a polycrystalline BeO inclusion in a polycrystalline Be matrix, the computed stress coefficients obtained at the inclusion matrix interface in accordance

with Equations 6 are approximately one order of magnitude greater
than that obtained for the pure Be system. The Table shows that
plastic yielding of the Be matrix material surrounding BeO inclusions
is a very probable event for any modest change in temperature.

Table 3.   Thermal Microstresses in Be and Be - BeO Systems.

	MPa/K		
Single Crystal Be Inclusion in Polycrystalline Be Matrix.	$\frac{\sigma_{11}}{\Delta T} = -0.07$		$\frac{\sigma_{33}}{\Delta T} = 0.15$
Polycrystalline BeO Inclusion in Single Crystal Be Matrix.	$\frac{\sigma_{11}}{\Delta T} = 1.74$		$\frac{\sigma_{33}}{\Delta T} = 1.43$
Polycrystalline BeO Inclusion in Polycrystalline Be Matrix.	$\frac{\sigma_{rr}}{\Delta T} = 1.14$		$\frac{\sigma_{\theta\theta}}{\Delta T} = -0.57$

PLASTIC FLOW AND FRACTURE

   Hill[10] has given an analysis of the extent of yielding to be
expected on a Tresca basis for an elastic-perfectly plastic material shell containing a central elastic inclusion. The value of
the compressive stress $\sigma_{rr} = -P$ at any position, r, beginning from
the inclusion interface and moving outward is connected with the
radial extent of the plastic zone size, c, by the relationship

$$P/P_o = 1 + 3 \ln(c/r) \tag{7}$$

where $P_o = (2 Y/3)$ and Y is the yield stress of the material. At
the matrix-inclusion interface P_a was determined by Hill as

$$P_a \approx (2 Y/3) \left[1 + \ln \left\{ E/3(1-\nu) Y \right\} \right] \tag{8}$$

where E is Young's modulus for the matrix and ν is Poisson's
ratio.

Figure 3 shows a graphical analysis of Hill's description of this problem. For the elastic case, straight lines of slope equal to - 1/3 are obtained for all values of induced stress at the matrix-inclusion interface, as obtained in the description by Gurland[7], and this should hold up to values of $P/P_o = 1.0$. At larger values of $-\sigma_{rr} = P > P_o$, yielding will be initiated at the matrix-inclusion interface and, depending on how large P might

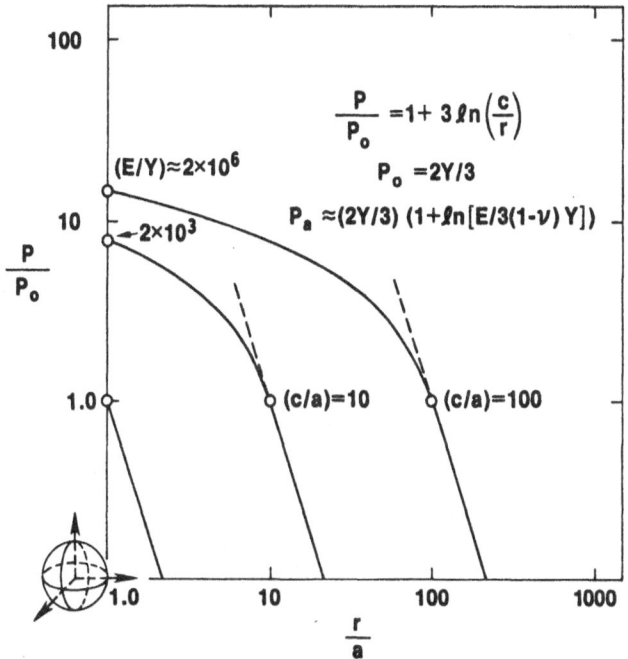

Figure 3.   Elastic – perfectly plastic analysis of radial stress for an inclusion in hydrostatic compression.

increase to, an increasing plastic zone radius is achieved, say,
even one equal to ten times the inclusion radius as indicated in
Figure 3. This spread of yielding is indicated in Figure 3 to be
reached at a value of $P_a \approx 7.9\ P_o$ according to Equation 7 and this
corresponds to an approximate ratio of Young's modulus and the
yield stress of 2×10^3, according to Equation 8. The very great
extent to which plastic yielding can prevent the build-up of large
elastic stress within the matrix is evident in Figure 3 from the
extrapolation of the dashed elastic curve. A plastic zone radius
100 times larger than the inclusion radius is estimated to be
obtained for a 1000 times greater ratio of Young's modulus and
yield stress than has been described in the previous case.

A Young's modulus-yield stress ratio of 2×10^3 is a typical
value for Be material and for other materials also. A yield
stress value corresponding to this ratio is easily achieved for
those materials with the lowest yield stress values described
previously by Armstrong and Borch[1]. On the scale of several
grains in Be it can easily be imagined that even lower values of
yield stress should apply because of the low shear strengths
expected for individual or even small groups of partially con-
strained crystals of a hexagonal metal. Thus, the cracking
description in Figure 2 seems to be a reasonable possibility of
the thermal microstresses in Be materials.

Liu and Gurland[11] have measured by x-ray diffraction experiments thermally-induced residual stresses in the silicon phase of
aluminum-silicon alloys of different compositions. For the eutectic
composition of 13.2% Si by volume, they measured the sum of compressive biaxial stresses equal to -22.4 MPa. This corresponds in
the current analysis to a value of $P = 11.2$ MPa. A value of $Y = 8.4$ MPa was reported for the alloy. This gives an estimated value
of $P/P_o \approx 2.0$ and, therefore, a plastic zone radius on the order
of 1.4 times the silicon particle radius is estimated from Equation
7 to occur. With $E = 7 \times 10^3$ MPa and $\nu = .34$ for Al, a plastic
zone radius on the order of 7.5 times the silicon particle radius
is estimated from Equation 8. Although the difference in radius
values estimated for the plastic zone is fairly large, it seems
important that plastic yielding is concluded to occur about the Si
particles on either basis of estimation. The neglect of any work
hardening behavior in the model calculations and the clustering of
the silicon particles in the Al-Si experimental measurements might
have contributed something to the difference in estimated results.

SUMMARY

Inclusions of BeO in Be appear quite capable of causing
plastically yielded zones and microcracking of the surrounding

matrix Be material. This consideration follows from previous work giving emphasis to the importance of thermal stresses in pure Be material due to its own elastic and thermal anisotropy as well as from the application to a BeO-Be composite system of Hill's elastic-perfectly plastic analysis for the internal stress values and corresponding plastic zone sizes. An analysis of previous experimental measurements of residual stresses in the eutectic Al-Si system, by Liu and Gurland, indicates that plastic yielding should have occurred around the Si particles in this system also.

REFERENCES

1. R. W. Armstrong and N. R. Borch, Thermal Microstresses in Beryllium and Other HCP Materials, Metallurgical Transactions. 2:3073 (1971).

2. L. E. Pope and A. L. Stevens, Wave Propagation in Beryllium Single Crystals, in: "Metallurgical Effects at High Strain Rates," R. W. Rohde, B. M. Butcher, J. R. Holland, and C. H. Karnes, eds. Plenum Publishing Corporation, New York (1973); p. 349.

3. S. J. Burns, J. Gurland, and M. H. Richman, Application of Fractographic Techniques to Beryllium, Metallography. 4:533 (1971); R. W. Armstrong, S. J. Burns, J. Gurland, and M. H. Richman, Internal Structural Factors Determining the Flow and Fracture Strengths of Beryllium, Air Force Materials Laboratory Technical Report. AFML-TR-69-10 (1969).

4. G. Simmons and H. Wange, "Single Crystal Elastic Constants and Calculated Aggregate Properties," 2nd edition. M. I. T. Press, Cambridge Mass. (1971).

5. Y. S. Touloukian, R. K. Kirby, R. E. Taylor, and T. Y. R. Lee "Thermal Expansion - Nonmetallic Solids", Plenum Publishing Corporation, New York (1977); and, "Thermal Expansion - Metallic Elements and Alloys", Plenum Publishing Corporation (1975).

6. V. A. Likhachev, Microstructural Strains Due to Thermal Anisotropy, Soviet Physics - Solid State. 3:1330 (1961).

7. J. Gurland, Temperature Stresses in the Two-Phase Alloy WC-Co, Trans. Am. Soc. Metals. 50:1063 (1958).

8. D. Burgreen, "Elements of Thermal Stress Analysis", 1st ed., C. P. Press, Jamaica, N. Y. (1971).

9. A. N. Stroh, A Theory of the Fracture of Metals, Advances in Physics. 6:418 (1957); R. W. Lardner, "Mathematical Theory of Dislocations and Fracture", University of Toronto Press (1979); p. 216.

10. R. Hill, "The Mathematical Theory of Plasticity", Clarendon Press, Oxford (1950).

11. C. T. Liu and J. Gurland, Thermally Induced Residual Stresses in Silicon Phase of Al-Si Alloys, Trans. Am. Soc. Metals. 58:66 (1965).

ACKNOWLEDGMENT

    The authors thank Professor Joseph Gurland, Brown University, for supplying Figure 1.

    This research has been completed as part of the Ph.D. thesis requirement intended to be submitted by T. A. Hahn to the Graduate School of the University of Maryland.

APPENDIX

LIST OF CONTRIBUTIONS TO THE SYMPOSIUM FOR
WHICH MANUSCRIPTS WERE NOT PUBLISHED

APPENDIX: LIST OF CONTRIBUTIONS TO THE SYMPOSIUM FOR
WHICH MANUSCRIPTS WERE NOT PUBLISHED

SESSION 1: HIGH AND LOW TEMPERATURE MEASUREMENTS

THERMAL EXPANSION OF SILICON-BASE GAS TURBINE CERAMICS

D. C. Larsen and H. H. Nakamura
IIT Research Institute

R. Ruh
U.S. Air Force Materials Laboratory

SESSION 2: MEASUREMENT TECHNIQUES

AUTOMATIC MULTISPECIMEN DILATOMETER

P. S. Gaal
Anter Laboratories

DEGREE OF CURE AND THE RUBBERY COEFFICIENT OF LINEAR
EXPANSION OF NITRILE-BUTADIENE RUBBER

D. L. Shelley
Armstrong Cork Co.

SESSION 5: THEORY AND CORRELATIONS

NON-LINEAR MODELING OF THERMAL EXPANSION DATA

R. K. Kirby
U.S. National Bureau of Standards

ELASTIC STIFFNESS AND THERMAL EXPANSION COEFFICIENT
OF (SILICIDE AND DIELECTRIC) THIN FILMS

A. K. Sinha and T. F. Retajczyk
Bell Laboratories

SESSION 6: MISCELLANEOUS MATERIALS/APPLICATIONS

THE THERMAL EXPANSION OF IMPORTANT HIGH EXPLOSIVES; ANISOTROPY
OF NITROGUANIDINE AND SYM-TRIAMINOTRINITROBENZENE

J. R. Kolb
Lawrence Livermore Laboratory

INDUSTRIAL APPLICATIONS OF MODERN DILATOMETRY IN
THE TEMPERATURE RANGE 5-2000 K

U. Hädrich and E. Kaiserberger
NETZSCH Gerätebau GmbH

H. Pfaffenberger
NETZSCH Brothers, Inc.

THERMAL LINEAR EXPANSION OF FOUR SELECTED AISI STAINLESS STEELS

P. D. Desai
CINDAS/Purdue University

STANDARD REFERENCE MATERIALS FOR THERMAL EXPANSION
CERTIFICATION OF STAINLESS STEEL AND A NEW STOCK OF COPPER

T. A. Hahn
U.S. National Bureau of Standards

LIST OF AUTHORS

SUBJECT INDEX

Amorphous Invar alloys, 163
Anisotropy, 29
Anomaly in expansion, 163
Antiferromagnetic transition, 29
Austenite to martensite
    transformation, 68

Composites, graphite-reinforced
    glass, 89
Cryogenic temperature, 67
Curie temperature, 163

Discontinuity in expansion upon
    melting, 131

Effect of cold-working, 163, 185
Exfoliation in graphite, 37
Expansion measurement
    dilatometer, 3, 45, 67, 89
    gamma attenuation, 17
    laser interferometer, 55, 103
    strain gages, 29
    thermomechanical analysis, 37

Ferrous fluoride (FeF_2), 29
Figures of merit, 121
Fluid saturated rocks, 173
Framework silicates
    and nitrides, 147

Graphite intercalation compound,
    37

Interferometric dilatometer for
    production, 55
International Thermophysics
    Congress, 85

Magnetic properties, 185
Melting alkali halides, 131
Microcracking in ZrO_2, 3
Microstresses in Be, 195
Molten nickel, 17

Neel temperature, 29

Phase transformation
    in FeF_2, 29
    in Fe-Ni-C, 67
    in MnCu, 45
    in ZrO_2, 3
Prediction of thermal expansion,
    139
Pressure dependence, 147, 157

Thermal expansion
    of alkali halides, 131, 139
    of β-spodumene, 103
    of Fe-B amorphous alloys, 163
    of FeF_2 crystals, 29
    of Fe-Ni-C alloys, 67
    of Fe-Pd Invar alloys, 185
    of glass-matrix composites, 89
    of graphite, 37
    of Invar, 103
    of mercurous chloride, 157
    of MnCu alloys, 45
    of Nb_2O_5, 103
    of nickel, 17
    of porous rocks, 173
    of zirconia, 3
Thermal stress resistance, 121

Zero CTE materials, 103

212